The carbon footprint
of everything

Mike Berners-Lee

P
PROFILE BOOKS

First published in Great Britain in 2010 by
PROFILE BOOKS LTD
3A Exmouth House
Pine Street
London EC1R 0JH
www.profilebooks.com

10 9 8 7 6 5

Printed and bound in Great Britain by
CPI Bookmarque, Croydon, Surrey

A CIP catalogue record for this book is available from the British Library.

ISBN 978 1 84668 891 1
eISBN 978 1 84765 182 2

Mixed Sources
Product group from well-managed
forests and other controlled sources
www.fsc.org Cert no. TT-COC-002227
© 1996 Forest Stewardship Council

'If we're serious about really addressing climate change, we need to become energy and carbon literate, and get to grips with the implications not only of our choices but also the bigger infrastructures which underpin the things we consume. How can we educate our desires unless we know what we're choosing between? Mike Berners Lee, to my complete delight, has provided just the wonderful foundation we need – a book that somehow made me laugh while telling me deeply serious things.'

Peter Lipman, Director of SUSTRANS

'Enjoyable, fun to read and scientifically robust. A triumph of popular science writing.'

Chris Goodall, author,
Ten Technologies to Fix Energy and Climate

'Curiously fascinating to both climate geeks and well-rounded human beings alike.'

Franny Armstrong, Director of *The Age of Stupid* and founder of 10:10

700039221559

lirector of
company
cialises in
e change.

Contents

Acknowledgements   ix

Introduction   xi

A quick guide to carbon and carbon footprints   1

Under 10 grams   11

A text message   *11*

A pint of tap water   *12*

A web search   *12*

Walking through a door   *14*

An email   *15*

Drying your hands   *17*

A plastic carrier bag   *18*

10 *to* 100 grams   21

A paper carrier bag   *21*

Ironing a shirt   *22*

Cycling a mile   *23*

Boiling a litre of water   *24*

An apple   *26*

A banana   *27*

An orange   *29*

An hour's TV   *30*

100 grams *to* 1 kilo   35

A mug of tea or coffee   *35*

A mile by bus   *37*

A nappy   *38*

A punnet of strawberries   *39*

A mile by train	40
A 500 ml bottle of water	43
A letter	44
1 kg of carrots	46
A newspaper	47
A pint of beer	49
A bowl of porridge	51
A shower	52
An ice cream	53
A unit of heat	54
A unit of electricity	55
Spending £1	59
1 kg of rubbish	61
Washing up	63
A toilet roll	64
Driving 1 mile	65
A red rose	68
1 kg of boiled potatoes	69
A pint of milk	71
1 kg of cement	74
1 kilo *to* 10 kilos	**75**
A paperback book	75
A loaf of bread	77
A bottle of wine	78
1 kg of plastic	80
Taking a bath	81
A pack of asparagus	83
A load of laundry	84
A burger	86
A litre of petrol	87
1 kg of rice	89
Desalinating a cubic metre of water	91
A pair of trousers	93
A steak	95
A box of eggs	96

1 kg of tomatoes   97
1 kg of trout   99
Leaving the lights on   100
1 kg of steel   101

10 kilos *to* 100 kilos   105

A pair of shoes   105
1 kg of cheese   106
A congested commute by car   107
A night in a hotel   108
A leg of lamb   111
A carpet   112
Using a mobile phone   113
Being cremated   115

100 kilos *to* 1 tonne   117

London to Glasgow and back   117
Christmas excess   119
Insulating a loft   121
A necklace   123
A computer (and using it)   124
A mortgage   127

1 tonne *to* 10 tonnes   131

A heart bypass operation   131
Photovoltaic panels   133
Flying from London to Hong Kong return   135
1 tonne of fertiliser   138
A person   139

10 tonnes *to* 100 tonnes   141

A car crash   141
A new car   143
A wind turbine   146
A house   149

100 tonnes *to* 1 million tonnes   **151**

Having a child   *151*
A swimming pool   *152*
A hectare of deforestation   *154*
A space shuttle flight   *155*
A university   *156*

1 million tonnes and beyond   **159**

A volcano   *159*
The World Cup   *160*
The world's data centres   *161*
A bushfire   *162*
A country   *164*
A war   *169*
Black carbon   *170*
The world   *171*
Burning the world's fossil fuel reserves   *175*

More about food   **177**

How the footprint of food breaks down   *177*
Low-carbon food tips   *182*
A guide to seasonal food   *183*

Some more information   **187**

Some assumptions revisited   *187*
The cost efficiency of selected carbon-saving options   *191*
Where the numbers come from   *193*
Carbon tables for countries, people, industries and products   *197*

Notes and references   **205**
Index   **229**

Acknowledgements

My biggest thanks go to Liz, Bill and Rosie for brilliant support and understanding, especially over Christmas when the deadlines loomed largest.

The book could not have happened without Duncan Clark, Mark Ellingham and others at Profile. Duncan's edits and advice have been superb throughout the project.

Many thanks to Jess Moss at Small World for unearthing quirky data and sorting out dozens of references – and for reminding me to get a move on. David Howard, Kim Kaivanto, Andy Scott and Geraint Johnes from Lancaster University and Sonny Khan all helped with the input-output model that I have drawn upon extensively. Thanks also to David Parkinson and Chris Goodall, among others, for answering technical queries.

Andrew Meikle let me chatter away during lift shares and has been a frequent sounding board. He read early pages aloud so I could hear how bad they were. Others who cast a friendly eye include Phil and Jane Latham, Aly Purcell, Rachel Nunn and Mark Jameson. Mum and Dad, true to form as incredible parents, both picked through the entire draft at a moment's notice.

Kim Quazi helped me thrash out the first ideas years ago in a pub. Going even further back the list of people to be grateful to is clearly endless but I want to thank Lee Pascal, David Brazier and Simon Loveday for three very different contributions.

I'm grateful to many of Small World's clients for providing material, but especially to Booths supermarkets, Lancaster University, the Crichton Carbon Centre, Historic Scotland and the Keswick Brewing Company.

Finally, thanks to everyone who said 'Oh, you're writing a book … how interesting!' and to those who, just to keep my morale up, pre-ordered copies long before I'd even finished the first draft.

Introduction

A few years ago I agreed to go round a supermarket with a journalist who wanted to write an article on low-carbon food. We trailed up and down the aisles with the dictaphone running and she plied me with questions, most of which I was pitifully unable to answer.

'What about these bananas? ... How about this cheese? ... It's organic. That must be better ... isn't it? ... Or is it? ... Lettuce must be harmless, right? ... Should we have come here by bus? ... At least we didn't fly! How big a deal is food anyway?'

It was not at all clear what the carbon-conscious shopper should do. There was clearly a huge gap in the available consumer knowledge and on that day we couldn't fill it. The article never happened, and it's probably just as well. Since then, I have looked long and hard into all kinds of carbon footprints, and carried out numerous studies, including one for a supermarket chain.

This book is here to answer the journalist's questions, and many more besides. It's not just a book about food and travel. I want to give you a sense of the carbon impact – that is, the climate change impact – of everything you do and think about. I want to give you a **carbon instinct**. Although I have discussed the footprint of just under one hundred items, I hope by the time you have read about these you will have gained such a sense of where carbon impacts come from that you will be able to make a reasonable guesstimate of the footprint of more or less anything and everything that you come across. It won't be exact, but I hope you'll at least be able to get the number of zeros right most of the time. There are messages here for personal lives, for businesses and a few sprinkled in for policy makers too.

Some basic assumptions

I'm hoping I can take three things for granted:

- climate change is a big deal;
- it's man-made
- and we can do something about it.

However, out of respect for the still widespread confusion over these assumptions, I have put more about them in an appendix in case you want to check them out before moving on.

Perspective

A friend recently asked me how he should best dry his hands to reduce his carbon footprint; with a paper towel or with an electric hand drier. The same person flies across the Atlantic literally dozens of times a year. A sense of scale is required here. The flying is tens of thousands of times more important than the hand drying. So my friend was simply distracting himself from the issue. I want to help you get a feel for roughly how *much* carbon is at stake when you make simple choices – where you travel to, how you get there, whether to buy something, whether to leave the TV on standby and so on.

Picking battles

I'm not trying to give you a list of 500 things you can do to help save the planet.[1] You could probably already write that list yourself. You will find at least 500 possibilities in here, but this is a book about helping you work out where you can get the best return for your effort. This book is here to help you **pick your battles**. If you enjoy the read and by the end of it have thought of a few things that can improve your life while cutting a decent chunk out of your carbon, then I'll be happy. The book isn't here to tell you what to do or how radical to be. Those are personal decisions.

Is carbon like money?

In one sense, yes it is.

Carbon is just like money in that you can't manage it unless you understand it, at least in broad terms. Most of the time we know how much things cost without looking at the price tag. I don't mean that we have an exact picture, but we know that a bottle of champagne is more expensive than a cup of tea but a lot cheaper than a house. So most of us don't buy houses on a whim. Our financial sense of proportion allows us to make good choices. If I really want champagne I know I can have it, provided that somewhere along the line I cut out something just as expensive that is less important to me. Our carbon instinct needs to be just like the one we have for managing our money.

That's where the similarity ends. Unlike with money, we are not used to thinking about carbon costs. It's also much harder to tell how much we are spending because we can't see it and it's not written down. Furthermore, unlike what happens when we spend a lot of money, we don't personally experience the consequences of our carbon impact because it's spread across nearly seven billion people and many years.

Enjoy the read

These pages are written for people who want to love their lives and for whom that now entails having some carbon awareness alongside everything else that matters to them.

Dip in. Keep it by the loo. Read it from cover to cover or flit around. Use it as a reference if you like. Talk about it. Take issue with it. Let me know how it could be improved (**info@howbadarebananas. com**). Think of it like an early map, full of inaccuracies but better, I hope, than what you had before.

If there's a fourth premise behind the book, it is that nearly all of us, including me, have plenty of junk in our lives that contributes nothing at all to the quality of our existence. It's deep in our culture. Cutting that out makes everyone's life better, especially our own. I got a big win by swapping my solo car commutes for bike rides and lift shares. That works for me, but I'm not prescribing that particular solution for you because we are all different. I hope you enjoy the read and that while you are at it you bump into at least something you can use.

So how bad are bananas?

As it happens, they turn out to be a fine low-carbon food though not totally free from sustainability issues to keep an eye on: see page 27.

A quick guide to carbon and carbon footprints

Carbon footprint is a lovely phrase that is horribly abused.[1] I want to make my definition clear at the outset.

Throughout this book, I'm using the word **footprint** as a metaphor for the total impact that something has.

And I'm using the word **carbon** as shorthand for all the different global-warming greenhouse gases.

So, I'm using the term **carbon footprint** as shorthand to mean the *best estimate* that we can get of the *full climate change impact* of something. That something could be anything – an activity, an item, a lifestyle, a company, a country or even the whole world.

CO_2e? What's that?

Man-made climate change, also known as global warming, is caused by the release of certain types of gas into the atmosphere. The dominant man-made greenhouse gas is carbon dioxide (CO_2), which is emitted whenever we burn fossil fuels in homes, factories or power stations. But other greenhouse gases are also important. Methane (CH_4), for example, which is emitted mainly by agriculture and land-fill sites, is 25 times more potent per kilogram than carbon dioxide. Even more potent but emitted in smaller quantities are nitrous oxide (N_2O), which is about 300 times more potent than carbon dioxide and released mainly from industrial processes and farming, and refrigerant gases, which are typically several thousand times more potent than carbon dioxide.

In the UK, the total impact on the climate breaks down like this: carbon dioxide (86 per cent), methane (7 per cent), nitrous oxide (6 per cent) and refrigerant gases (1 per cent).

Given that a single item or activity can cause multiple different greenhouse gases to be emitted, each in different quantities, a carbon footprint if written out in full could get pretty confusing. To avoid this, the convention is to express a carbon footprint in terms of **carbon dioxide equivalent (CO_2e)**. This means the total climate change impact of all the greenhouse gases caused by an item or activity rolled into one and expressed in terms of the amount of carbon dioxide that would have the same impact.[2]

Beware carbon toe-prints

The most common abuse of the phrase carbon footprint is to miss out some or even most of the emissions caused, whatever activity or item is being discussed. For example, many online carbon calculator websites will tell you that your carbon footprint is a certain size based purely on your home energy and personal travel habits, while ignoring all of the goods and services you purchase. Similarly, a magazine publisher might claim to have measured its carbon footprint but in doing so looked only at its office and cars while ignoring the much greater emissions caused by the printing house that produces the magazines themselves. These kinds of carbon footprint are actually more like carbon 'toe-prints' – they don't give the full picture.

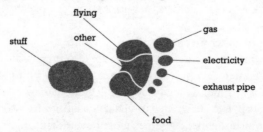

Figure 1.1. The footprint of a lifestyle is bigger than its toe-print.

Direct and indirect emissions

Much of the confusion around footprints comes down to the distinction between 'direct' and 'indirect' emissions. The true carbon footprint of a plastic toy, for example, includes not only direct emissions resulting from the manufacturing process and the transportation of the toy to the shop: it also includes a whole host of indirect emissions, such as those caused by the extraction and processing of the oil used to make the plastic in the first place. These are just a few of the processes involved. If you think about it, tracing back all the things that have to happen to make that toy leads to an infinite number of pathways, most of which are infinitesimally small. To make the point clearly, let's try following just one of those pathways. The staff in the offices of the plastic factory used paper clips made of steel. Within the footprint of that steel is a small allocation to take account of the maintenance of a digger in the iron mine that the steel originally came from ... and so on for ever. The carbon footprint of the plastic toy includes the lot, so working it out accurately is clearly no easy task!

To give another example, the true carbon footprint of driving a car includes not only the emissions that come out of the exhaust pipe, but also all the emissions that take place when oil is extracted, shipped, refined into fuel and transported to the petrol station, not to mention the substantial emissions caused by producing and maintaining the car.

A note about high-altitude emissions

Emissions from planes in the sky are known to have a greater impact than those that would arise from burning the same amount of fuel at ground level. The science of this is still poorly understood. Nevertheless, because our measure is setting out to be a guide to climate change impact it is essential to try to take this into account. That is why in this book I have multiplied all aviation emissions by 1.9.[3] (Some experts believe the true impact of plane emissions could be even higher, and suggest a multiplier of up to 4.)

In the Some more information section you'll find a slightly more technical discussion of the methodologies I have used to get the numbers (page 187).

The essential but impossible measure

The carbon footprint, as I have defined it, is *the* climate change metric that we need to be looking at. The dilemma is that it is also impossible to measure. We don't stand a hope of being able to understand how the impact of our bananas compares with the impact of all the other things we might buy instead unless we have some way of taking into account the farming, the transport, the storage and the processes that feed into those stages. A key question, then, is this: 'How should we deal with a situation in which the thing we need to understand is impossibly complex?'

One common response is to give up and measure something easier, even if that means losing most of what you are interested in off the radar. The illusionist Derren Brown refers to one of his core techniques as the *misdirection of attention*: by focusing his audience on something irrelevant he can make them miss the bit that matters. Examples include an airport waxing lyrical about the energy efficiency of its buildings without mentioning the flights themselves. The same thing can happen by accident. If you settle for a toe-print, there is a very good chance it will *misdirect* your attention away from the big deals.

An alternative response to the dilemma, and the approach that this book is all about, is to do the best job you can, despite the difficulties, of understanding the whole picture. This book is about making the most realistic estimates that are possible and practical, and being honest about the uncertainty.

Blurry numbers ...

First and foremost, I am trying to get the orders of magnitude clear.

In my work I put a lot of effort into developing a realistic picture of different carbon footprints using a variety of methods. This book draws upon a lot of that, as well as the most credible secondary sources that I have been able to find. However, huge uncertainty remains. So when you see a number like '2.5 kg CO_2e' on an item such as a burger, bear in mind that it is a best estimate. What it really means is something like 'best estimate of 2.5 kg CO_2e, probably between 1.5 and 4 kg CO_2e and almost certainly between 1 and 10 kg'. That is the nature of all carbon footprints. Don't let anyone tell you otherwise.

Some of the numbers you'll see are even flakier still. This generally happens when I'm trying to bring the beginnings of a sense of scale to important questions that are almost impossible to quantify. Sometimes my calculations and assumptions are highly debatable but I've included them because I think that just going through the thought process can be a useful reflection on something that matters. Examples include the footprint of having a child, a nuclear war or a text message.

If you think you can offer an improvement on any of the numbers in this book, I'll be very happy to hear from you.

... but they will do ...

Let me be emphatic that the uncertainty does not negate the exercise. Real footprints are *the essential measure* and nothing short of them will do. The level of accuracy that I have described is good enough to separate out the flying from the hand drying. And even if you use the numbers here to make finely balanced decisions, most of the time those choices will be better informed than if you had no guidelines at all.

... for now

That we find footprinting tricky is a problem for us all. The situation we are in is like sailing round the world with a map from the 1700s. How should we respond? Throw that map away and have nothing? Definitely not! Use a high-quality map of just a small part of the ocean and ignore the rest? No way. Use the maps we have but treat them with caution? Absolutely. Try to make better maps? Of course – and the work is ongoing. This book is just an early map. Better ones will follow. And this book is trying to help you improve the carbon map that you carry around in your own head.

Making sense of the numbers

So far we've established what we need to try and measure, but a tonne of carbon is still a highly abstract concept. I'm now going to try to give it a bit more real-life meaning.

What does a tonne of CO_2e look like?

If you filled a couple of standard-sized garden water butts to the brim with petrol and set fire to them, about a tonne of carbon would be directly released into the atmosphere. (The carbon footprint of burning that petrol by driving is a bit more than that, for reasons explained later.) If you did the same with a pint milk bottle, that would release just over a kilogram of carbon dioxide, and if you burned a blob about the size of a chickpea, that would release about a gram.

1000 grams (g) = 1 kilogram (kg)
1000 kilograms = 1 tonne

How many tonnes do we each cause?

To give a quick sense of scale, the average UK person currently has an annual carbon footprint of around 15 tonnes. The Chinese and Malawians emit less but the Americans and Australians more. There is more detail on this later on. You get smaller numbers if you only include the obvious bits of your footprint such as household energy and travel or you miss out emissions on goods you buy that are manufactured overseas.

The 10-tonne lifestyle

I'm not here to set you a particular target or to make you feel guilty. How you decide to live is a personal choice that only you can make. I just want to help you understand carbon so that you can do whatever you decide to do with more knowledge.

However, to help get a sense of perspective I have adopted a 10-tonne lifestyle as another unit of measure for this book. I am going to refer to it from time to time, because it gives an alternative and sometimes clearer way of conceiving of those abstract kilograms and tonnes of CO_2e.

Apart from being a round number, there is not much that is particularly magic about a 10-tonne lifestyle – that is, a lifestyle causing 10 tonnes of CO_2e per year. It's certainly not a long-term sustainable target for everyone in the world: if everyone went in for 10-tonne living all over the globe, emissions would skyrocket by 40 per cent.

On the other hand, truly sustainable long-term targets aren't practical or helpful in the short term. For example, the UK has a target to cut carbon emissions by 80 per cent by 2050. If you apply this to the stuff we import as well as to the emissions within the country itself, that would take us down to around 3 tonnes per person per year. Some commentators think we'll need to go even lower. Ultimately, though, it's virtually impossible for an individual in the developed world to get down to a 3-tonne lifestyle any time soon. That kind of cut requires the whole economy to be made greener.

Ten tonnes, by contrast, is a modest aspiration target that most people could meet with enough effort. In the UK and many other European countries, adopting a 10-tonne lifestyle would mean reducing your emissions to about one-third below average. In Australia and the US, it would mean a reduction of closer to two-thirds below average.

One way of thinking about the footprint of an object or activity is to put it in the context of a year's worth of 10-tonne living. For example, a large cheeseburger, with a footprint of 2.5 kg CO_2e, represents about 2 hours' worth of a 10-tonne year. If you drive a fairly thirsty car for 1000 miles, that is 800 kg CO_2e, or a month's ration. If you leave a couple of the (now old-fashioned) 100-watt incandescent light bulbs on for a year, that would be another month used up. One typical return flight from London to Hong Kong burns up around 4.6 tonnes CO_2e. That is just under 6 months' ration in the 10-tonne lifestyle.

A short car commute, a daily cheeseburger, and some wasteful lighting habits could easily use up a quarter of the 10-tonne budget. Then if you also take the flight to Hong Kong, that would leave just 3 months' ration left in the 10-tonne budget for *everything* else that year: other food, heat, buying stuff, health care, use of other public services, your contribution to the maintenance of roads, any wars around the world that your government is involved in (like it or not) – *the lot*.

You might be wondering whether there are any better ways of spending this or any other sized budget than blowing most of it on burgers, commuting and flying. If that question is of interest, this book has been written for you.

How many tonnes for a life or a death?

I hope the comparisons so far have helped to make a tonne of carbon seem a bit more tangible. But let's see whether it's possible to get a handle on how much it might actually *matter*. Our species is good at understanding the direct, immediate and visible consequences of our actions. We are a lot less smart at grasping the consequences when they are dispersed across billions of people whom we will never meet. This might not have mattered when we lived in caves but it won't let us live well in a global society. Our impacts used to be local and visible. Today they are not. Perhaps we need to find it as shocking when we see dispersed suffering inflicted through needless carbon emissions as it would be to see the same suffering inflicted all in one place in front of our eyes by, let's say, a street stabbing.

I did some 'back of the envelope' sums and arrived at a figure of 150 tonnes CO_2e per climate change-related death. I've spelled out my calculations in the endnote that follows this sentence.[4] If you look it up and follow my sums, you'll see that I don't have even the beginnings of a rigorous argument to justify my figure. But it was an interesting thought process and one that, if you do decide to follow it, you might even find faintly plausible. Or you may think my line of thought is hopelessly unrealistic. And maybe you would be right. I was just playing with ideas. It is up to you to decide what meaning to take from them. For me, even a possibility of any realism in this line of thought throws up a challenge.

The 150 tonnes per life figure would mean that if your lifestyle had the footprint of the average UK citizen, one person would have to die from climate change somewhere in the world every 10 years. If you were to fly to Hong Kong and back 11 times first class – that would be another death.

How much would it be worth paying to save a tonne of carbon?

This is not going to be an easy question to answer. An unknown number of lives depends on our response to climate change, and even if we did know how many, it is not as if our society has a consistent approach, even in the very broadest of terms, to determining the kind of value that each one of those lives might have. So, putting a financial value on the saving of a tonne of carbon is going to be tough, to put it mildly. Nevertheless, it's a question worth pondering because unless we understand there to be real and tangible value in cutting emissions, we will simply never bother and, for better or worse, money has become our language for understanding value.

As I write, £12 per tonne is the maximum price of CO_2 that companies in the UK could have to pay.[5] Let's see what happens if we work on that £12 figure. With global emissions at 50 billion tonnes, does that mean that the world might be prepared to pay just 600 billion pounds to eradicate our emissions completely? Is that really all it's worth to us? That's about three-quarters of a per cent of global output in economic terms to have a miracle cure for climate change? Surely on this basis carbon is worth a lot more than £12 per tonne.

Let's see what £12 per tonne implies if you link it in to my estimate of 150 tonnes per death. That would put the value of a life at just £1800. The value of the world's population under this analysis is a mere £12 trillion, or about six times the Gross Domestic Product of the UK. My home town of Kendal has about 24,000 people. Would it really be a good deal to blow up everyone in it if it would liberate £43 million? This analysis places the value of the UK population at just £108 billion. In other words, the people living in the UK are valued at about 5 per cent of their GDP.

So how much should it be worth in financial terms to save a tonne of carbon? A great deal more than the £12, clearly!

Under 10 grams

A text message

0.014 g CO$_2$e one message
32,000 tonnes CO$_2$e all world's texts for a year

The biggest part of a text message's footprint is the power used by your phone while you type – and of course by your friend's phone while they read what you've written. If the two of you take a minute between you to type and read the message, and you each have phones that consume 1 watt of power when in use, the message's footprint will be about a hundredth of a gram. This figure takes into account the transmission of a 140-character message across the network.[1]

Around the world, about 2.5 trillion texts are sent every year.[2] Don't be fooled into thinking that the 32,000 tonnes footprint for this total is a big number. It isn't. 32,000 tonnes is about one ten-thousandth of a per cent of the world's carbon footprint. In other words, texting is not a big deal. It wouldn't even be a big deal if my numbers were out by a factor of a hundred.

Incidentally, as of 2008, nearly a quarter of all text messages were sent in China, and about a fifth in the Philippines, where they average an impressive 15 messages per day for each phone. The average North American phone sent just a couple of messages a day, whereas British phones manage six texts per handset.

In summary, we can relax about sending texts (but no spam, please).

A pint of tap water

0.14 g CO$_2$e one pint
14 kg CO$_2$e a year's tap water for a typical UK citizen

A year's supply for one person is the same as a 20-mile drive in an average car.[3] That includes drinking, washing, cleaning – the lot.

Unlike the bottled alternative, which has around 1000 times the impact (see page 43), cold tap water is not a major carbon concern for most people. Indeed, the provision and disposal of household water accounts for less than half a per cent of the UK's carbon footprint.[4] Climate change looks set to cause serious water stress in some places. In the UK as a whole it looks as though we are going to have plenty, even though some redistribution might be called for.

Interestingly, if our pint of tap water is poured down the drain, its footprint leaps almost fourfold to just over half a gram because it is more carbon intensive to treat waste water than to supply the water in the first place.[5] If the eventual fate of the drink is to be flushed down the loo along with another 6 litres, that takes the total to 4 g CO$_2$e.

Tap water itself is one thing. Heating it up is another matter, accounting for a decent chunk of the typical person's emissions (see page 24). See also Swimming pool, page 152, and Desalination, page 91.

A web search

0.2 g CO$_2$e Google's estimate for the energy used at their end
0.7 g CO$_2$e from an efficient laptop – a lower estimate
4.5 g CO$_2$e from a power hungry machine and making higher estimates of power used in the network

So that is between 2 and 14 seconds' worth of ten tonne living for a 30-second single search.

At the low end of the scale, I've started off with Google's estimate of 0.2 g CO_2e for the electricity they use at their end when you put in a single search enquiry.[6] Add to it just 30 seconds of machine time at your end on an efficient 20-watt laptop while you tap in the search, wait for the result and scan it for what you want. That's another 0.1 g, bringing the total so far to 0.3 g. Your local network and the servers that actually host the information you are digging for probably come to at least 50 per cent of the amount of power used by your machine, even if they are super-efficient, like your laptop,[7] so that takes us to 0.35 g. Wear and tear and depreciation of hardware throughout the whole system probably doubles this because of the emissions that are required in the manufacture of all that kit. That takes us to 0.7 g CO_2e for a single enquiry that might let you, say, find the location of the restaurant you're heading to.

On a more power-hungry desktop computer that uses 150 watts of power, your web search might burn through about 0.75 g CO_2e. If you apply the same mark-ups for networks and hardware, we get to a grand total of 4.5 g, with Google accounting for just 0.2 g of that.

One can search for information about the footprint of web searches. You'd find blogs and articles all coming up with different figures based on different assumptions and all including different things. Some look at multiple searches and therefore produce much higher headline figures.[8]

At the high end of my estimate, the activity of surfing clocks up a carbon footprint at about half the rate of the 10-tonne life. In other words, if you spent a whole year browsing the web non-stop you'd trigger about 5 tonnes of emissions. That sounds good until you remember that at the same time you might also be wearing clothes, keeping warm, burning calories, getting closer to your next need for medical attention, living in a building that needs periodic maintenance and so on. Even while you are sat at the machine, your browsing is just one part of your footprint.

Google is estimated to deal with 200–500 million enquiries per day. If we go with the top estimate, and the high-end figure for the footprint of a single search, Google searching accounts for 1.3 million tonnes CO_2e per year. That is a big number, but it is only about one forty-thousandth of our global footprint. We can probably relax about it. Reading the stuff we find is an altogether more carbon-hungry activity – see page 15.

Walking through a door

Zero CO_2e a normal household door on a summer's day
3 g CO_2e getting in through your front door on a cold winter's day
84 g CO_2e big electric doors opening into a large stairwell on a cold windy day

At the high end, that's a banana's worth of greenhouse gas every time you enter the building.

The entrance door of the building where I work has no manual option.[9] To get in you have to press a button and wait while two electric motors whir and double doors swing slowly open, creating a space 2 m wide by 2.5 m high. You enter a spacious stairwell with two large radiators. The only decoration is a certificate proclaiming the 'D'-rated energy performance of the building. It takes 18 seconds for the doors to finish closing. This three-year-old building was amazingly rated environmentally 'Excellent' in its BREEAM assessment.[10]

The power used by the electric motors themselves isn't the problem. They account for just 1 g CO_2e. The problem is the size of the space you have to open, the time it has to stay open for and the vast heated space that the doors open onto.

For this building there must have been lots of other options, such as manual doors that swing shut and can be opened singly, with an override button for disabled access. Rotating doors attached to turbines that generate electricity as you pass through have been tri-alled in Holland but sound like the kind of gimmick that can tarnish the reputation of the renewables industry.

In a typical home on a cold, blustery day, the numbers are more likely to come out at about 3 g, based on opening it by hand and closing it straight away.

An email

0.3 g CO_2e a spam email
4 g CO_2e a proper email
50 g CO_2e an email with long and tiresome attachment that you have to read

A typical year of incoming mail adds up to 135 kg CO_2e: over 1 per cent of the 10-tonne lifestyle and equivalent to driving 200 miles in an average car.

The annual figure provided here is for the typical business user and includes the sending, filtering and reading of every incoming message. According to research by McAfee, a remarkable 78 per cent of those incoming emails are spam. Around 62 trillion spam messages are sent every year, requiring the use of 33 billion units of electricity and causing around 20 million tonnes of CO_2e per year. McAfee estimated that around 80 per cent of this electricity is consumed by the reading and deleting of spam and the search-ing through spam folders to dig out genuine emails that ended up there by accident. Spam filters account for 16 per cent. The actual generation and sending of the spam is a very small proportion of the footprint.

Although 78 per cent of incoming emails sent are spam, these mes-sages account for just 22 per cent of the total footprint of your email account because, although they are a pain, you deal with them quickly. Most of them you never even see. A genuine email has a bigger carbon footprint, simply because it takes time to deal with. So if you are someone who needlessly copies people in on messages just to cover your own back, so you can claim they should have known about it, the carbon footprint gives you one more good reason for changing your ways. You may find that after a while everyone at work starts to like you more, too.

The average email has just one-sixtieth the footprint of a letter (see page 44). That looks like a carbon saving unless you end up sending 60 times more emails than the number of letters you would have posted in days gone by. Lots of people do. This is a good example of the *rebound effect* – a low-carbon technology resulting in higher-carbon living simply because we use it more.

If the great quest is for ways in which we can improve our lives while cutting carbon, surely spam and unnecessary email have to be very high on the hit list along with old-fashioned junk paper post.

If only email were taxed. Just a penny per message would surely kill all spam instantly. The funds could go to tackling world poverty, say. The world's carbon footprint would go down by 20 million tonnes even if genuine users didn't change their habits at all. The average user would be saved a couple of minutes of their time every day and there would be a £170 billion annual fund made available. If 1p turned out to be enough to push us into a more disciplined email culture – with perhaps half the emails sent – the anti-poverty fund would be cut in half but a good few minutes per day would be liberated in many people's lives and the carbon saving would be around 70 million tonnes CO_2e – that's nearly as much as all UK household electricity.

Drying your hands

Zero CO$_2$e letting them drip
3 g CO$_2$e Dyson Airblade
10 g CO$_2$e one paper towel
20 g CO$_2$e standard electric drier

On average, if you used public toilets six times per day, your hand drying would produce around 15 kg per year; equivalent to 1 kg of beef.

'What's the greenest way to dry my hands?' is a frequently asked question, so I'll answer it even though I have already made the point that if you really want a lower-carbon lifestyle you should be asking about something more important.

Close to the low end of the scale is drying your hands with a Dyson Airblade. This dryer does the job in about 10 seconds with 1.6 kilowatts of power. Its secret is that it doesn't heat the air. It just blows it hard. This makes it far more efficient than conventional hand driers.

In the middle of the spectrum I have put paper towels, based on 10 g of low-quality recycled paper per sheet, and only one towel used each time.[11] (Of course, if you use two or three towels the footprint doubles or triples.)

At the high end are conventional heated hand driers. These take a shade longer than the Dyson and use around 6 kilowatts of power. The big difference is explained by the fact that it always takes a lot of energy to create heat.

Right at the bottom of the scale comes not drying your hands at all – or indeed using a small hand towel that is reused many times in between low-temperature washes. I am not a hygiene expert but I'm told that neither option is good from that point of view: they may even end up adding to the already substantial footprint of the health service (see page 13).

A plastic carrier bag

3 g CO_2e very lightweight variety
10 g CO_2e standard disposable supermarket bag[12]
50 g CO_2e heavyweight, reusable variety

So that's 2.5 kg per year if you use five standard bags per week: about the same as one large cheeseburger.

Over the past few years the UK supermarkets have been making a big effort to reduce the use of plastic bags. It's a highly visible green gesture and as such I'm not criticising it. But has it helped us get to grips with climate change? Not really: when someone in the developed world walks home from the shops with a disposable plastic bag full of food, the bag is typically responsible for about one-thousandth of the footprint of the food it contains. In other words, it is good if your supermarket is taking action on plastic bags, but don't let that stop you from asking what it is doing about the other 999 thousandths of its carbon agenda.

Carbon emissions are not the only environmental problem associated with plastic bags, of course. They also have a habit of hanging around in the ecosystem where they can sit for hundreds of years, clogging up animals' stomachs, killing fish and being ugly. *National Geographic* estimates that the world uses between 500 billion and 1 trillion disposable grocery bags per year.[13] That's an awful lot of rubbish – even if the bags contribute only around one ten-thousandth of the world's total carbon footprint.

How best to get rid of them, then? Burning releases nasty toxins as well as carbon, although the technology is improving. From a purely climate-change perspective, landfill is not too bad. They won't degrade, so all those hydrocarbons are returned to the ground where they came from for fairly long-term storage. But landfill is nasty for other reasons.

So, although disposable plastic bags aren't a serious carbon issue, it does make sense to avoid using them where possible. Better alternatives are a rucksack (which makes things easier to carry and keeps your hands free), a wheelie basket (which avoids you having to lift things at all) or sturdy, reusable bags. If you do use reusable plastic bags, make sure you really do reuse them: if you get less than five uses out of one, you'd be better off, in carbon terms, with disposable ones.

10 grams *to* 100 grams

A paper carrier bag

12 g CO$_2$e recycled and lightweight
80 g CO$_2$e an elaborate bag from mainly virgin paper as supplied by many clothing retailers

A common misconception is that paper bags must be lower carbon than plastic. Wrong! The paper industry is highly energy intensive. Printed virgin paper typically produces between 2.5 and 3 kg CO$_2$e per kilo of paper manufactured. This is comparable to the emissions required produce 1 kg of polypropylene plastic bags. However, paper bags have to be much heavier, so overall the paper bag ends up having a bigger footprint.

Recycled paper is roughly half as energy intensive to produce as virgin paper. But even a lightweight recycled paper bag produces slightly more greenhouse gas emissions than a typical plastic carrier.

There is another problem at the disposal end as well, which I have not factored into my numbers. Unless you recycle your paper bag, it is likely to end up in landfill, where it will rot, emitting more CO$_2$ and, even worse, methane. Landfill sites vary in their ability to capture and burn methane emissions, but typically there will be around 500 g of greenhouse gas emissions per kilo of paper buried.[1]

One final detail about paper bags is that they often don't work, resulting in bruised apples rolling down the street.

Low-carbon tips

■ If given a choice between plastic and paper, the plastic one may well be best (see page 18)

■ If stuck with paper, recycle it when you are done with it. (It is probably too much to hope that it could be fit for re-use.)

Ironing a shirt

14 g CO$_2$e a quick, expert skim on a slightly damp shirt
25 g CO$_2$e average
70 g CO$_2$e a thoroughly crumpled shirt ironed by unskilled hands.

Five shirts every week is about the same as a 10-mile drive once a year in an average car.

A friend of mine used to iron her husband's socks (she's now divorced). If you're feeling stuck in a similar routine, I hope you will find the carbon argument gives a bit more power to your elbow.

Although ironing isn't the biggest environmental issue, there may be scope for saving a little bit of carbon here – and perhaps some lifestyle improvement, too. For ironing that simply has to be done, the best green step is to have the clothes slightly damp and use the ironing process itself to finish off the drying. That saves both time and carbon (especially if you otherwise would be using an energy-hungry tumble drier, see page 84). Even more effective is simply using the iron less often.

A few people allegedly enjoy this activity, almost as a hobby. If ironing is how you get your kicks, it works out at about 400 g CO$_2$e per hour. That's about five times worse than watching the average TV but dramatically better than going for a drive. I have also heard ironing described as having meditative value. I can only assume that this goes something along the lines of 'a deep reflection on the resentment you notice inside yourself at spending your time in this way'. If this is you, can I recommend a good old-fashioned, Zen-style breathing routine, weighing in at zero g CO_2e?

Cycling a mile

65 g CO_2e powered by bananas
90 g CO_2e powered by cereals with milk
200 g CO_2e powered by bacon
260 g CO_2e powered by cheeseburgers
2800 g CO_2e powered by air-freighted asparagus

If your cycling calories come from cheeseburgers, the emissions per mile are about the same as two people driving an efficient car.

I have based all my calculations on the assumption that you burn 50 calories per mile.[2] The exact figure depends on how fit you are (the fitter you are, the lower the figure), how heavy you are, how fast you go (the faster, the higher) and how much you have to use the brakes.

All that energy has to come from the food you eat and that in turn has a carbon footprint. The good news is that the lower-carbon options are also the ones that make the best cycling fuel.

Bananas, of course, are brilliant (see page 27). Breakfast cereal is pretty good (let down slightly by the milk). The bacon comes in at around 200 g CO_2e for a 25 g rasher with only enough calories for a mile and a quarter of riding.

As mentioned above, two people cycling along using calories from cheeseburgers is equivalent to those same people sharing a ride in an efficient car. At the ridiculous high end of the scale, however, is getting your cycling energy by piling up your plate with asparagus that has been flown by air from the other side of the world. At 2.8 kg per mile this is like driving a car that does six miles to the gallon (a shade over a mile per litre). You'd be better off in a Hummer.

All my figures include 50 g per mile to take into account the emissions that are embedded in the bike itself and all the equipment that is required to ride it safely.[3] In the lower-carbon scenarios, the food accounts for only a small part of your impact, and the maintenance of bike and sundry equipment dominates.

Is cycling a carbon-friendly thing to do? Emphatically yes! Powered by biscuits, bananas or breakfast cereal, the bike is nearly 10 times more carbon efficient than the most efficient of petrol cars. Cycling also keeps you healthy, provided you don't end up under a bus. (Strictly speaking, dying could be classed as a carbon-friendly thing to do but needing an operation couldn't: see page 131).

Buying a folding bike so I could commute on the train has been one of the best decisions I have made in recent years – in terms of both lifestyle and carbon. My journey takes 10 minutes longer, but I get half an hour's exercise and 15 minutes reading a book each way. Because both of those are things I like doing but struggle to find enough time for, I've magicked an extra hour of the stuff I love into my day – while saving money and carbon.

One other thing: by taking my car off the road in rush hour, I cut everyone else's queuing time as well, and reduce the emissions they belch out while they wait (see Congested car commute, page 107).

Boiling a litre of water

50 g CO_2e gas kettle, fairly low heat
70 g CO_2e electric kettle
115 g CO_2e saucepan on the gas without a lid and flames up the side

Some friends of ours have a stove-top kettle that they use on their gas cooker, and we ended up debating the environmental pros and cons for months. Finally I spent half a morning measuring different methods. (A sad way of spending time, I know, but I did have a book to write.)

Our plug-in electric kettle was the fastest. Only 10 per cent of the electrical energy was wasted, so although inefficiencies in our power stations and distribution systems make electricity a high-carbon way of producing heat, the electric kettle is still a fairly good way of boiling water at home.

How the gas kettle compares with the electric kettle depends on the time of year. In winter, our friends win the low-carbon prize easily. That's because although some of the heat from the gas flames escapes around the edge of their kettle, that heat isn't actually wasted: the kitchen is the heart of their house, so all the heat that goes into the room is useful. In their house, in fact, the gas cooker is the most efficient form of heating because nothing is wasted up the flue (as it is with a gas boiler), nor is any heat sent to unoccupied rooms or lost in pipework (as it is with central heating).

In the summer, our friends still win the low-carbon prize provided they are willing to put their kettle on a small gas ring to maximise the proportion of the heat that goes into the water, rather than being lost around the sides. Doing this gives them a 30 per cent carbon saving over the electric kettle but also means it takes three times as long (12 minutes) to boil. If they use large gas ring, the result is slightly *more* carbon than the electric kettle – and it's still 50 per cent slower.

Saucepans turned out to be less efficient than kettles. It only makes sense to bring water to the boil in a saucepan if you are putting vegetables in at the start, in which case there is the benefit that they begin cooking a bit even before the water boils. If you do use a saucepan, keep the lid on (20 per cent waste if you don't) and make sure the flames don't go up the sides (potential for another 20 per cent waste).

To summarise, kettles are better than saucepans, and gas beats electric – but only if you are not in a hurry or you want to heat your room anyway. Just as important, of course, is not to boil more water than you actually need.

Four kettle design improvements are worth a mention since there are some incredibly simple features waiting to hit the mass market.

- Although only about 10 per cent of the heat generated by an electric kettle is wasted, I was surprised at how hard it was to find a kettle with proper insulation. Better insulation would also mean that if you forget it has boiled, or you accidentally boil more than you need, it would stay hot for longer.

- A thermostat so you can set it to 85°C when that is all you need – such as when making coffee or herbal tea. This is quicker, cheaper, lower carbon and probably reduces the chance of mouth cancer. The Morphy Richards Ecolectric Kettle is the only one I've found with this feature.

- An old-fashioned whistle or a beep option would stop you forgetting when it has boiled.

- The Eco Kettle, already on the market, allows you to decant just the amount you need from a reservoir, making it easier to boil only what you need.

The Dragons' Den must be waiting for someone to put all these features together.

An apple

Zero CO_2e plucked from the garden
10 g CO_2e local and seasonal
80 g CO_2e average; that's 550 g per kilo
150 g CO_2e shipped, cold stored and inefficiently produced

Apples are a low-carbon food wherever they come from. Beyond that it is difficult to be certain about the details.

One study from a university in New Zealand found that apples grown in that country for the UK market incurred just 185 g CO_2e per kilo – significantly lower than UK apples for local consumption, which came it at 271 g per kilo.[4] The argument made in the study was that UK production entailed greater use of fossil fuels on the farm and required more cold storage. The study also cited New Zealand's cleaner electricity mix. These factors, it claimed, outweighed the emissions from shipping the produce half way around the world. A similar comparative study referenced by the UK government's Department for Environment, Food and Rural Affairs (Defra) produced similar orders of magnitude but found, conversely, that for Germany (which should be similar to the UK) local apples were more carbon friendly than those sourced from New Zealand.[5] It's difficult to unpick the arguments and determine who got closer to the truth. Each study went about things slightly differently and made different assumptions. This story illustrates an important point: these kinds of study are always tricky, heaped with far more uncertainties and subjective judgements than many people like to admit.

As you'd expect, local, in-season apples are best but there is nothing particularly bad about buying them from anywhere in the world, because they travel on a boat rather than a plane. Indeed, in early summer, when any local apples will have been in cold storage for months, importing may be the lower-carbon option.

One last point: as with all fruit and vegetables, it's good idea to buy the most misshapen ones you can get, because that encourages the supply chain not to chuck them in the bin before they ever reach the shops.

A banana

80 g CO_2e each, or 480 g per kilo[6]

To answer the question in the title of this book, bananas aren't bad at all. They're brilliant! To emphasise the point, I'm eating one as I write.

Bananas are a great food for anyone who cares about their carbon footprint. For just 80 g of carbon, you get a whole lot of nutrition: 140 calories as well as stacks of vitamin C, vitamin B6, potassium and dietary fibre. Overall, they are a fantastic component of the low-carbon diet. Bananas are good for just about everyone – athletes, people with high blood pressure, everyday cycle commuters in search of an energy top-up, or anyone wishing to chalk up their recommended five portions of fruit and vegetables per day. There are three main reasons that bananas have such low carbon footprints compared with the nourishment they provide:

- They are grown in natural sunlight – no hot-housing required.

- They keep well, so although they are often grown thousands of miles from the end consumer, they are transported by boats (about 1 per cent as bad as flying).

- There is hardly any packaging, if any, because they provide their own.*

On top of their good carbon and healthy eating credentials, the fair trade version is readily available.[7]

For all their good qualities, don't let me leave you with the impression that bananas are too good to be true. There are environmental issues. Of the 300 types in existence, almost all those we eat are of the single, cloned 'Cavendish' variety. The adoption of this monoculture in pursuit of maximum, cheapest yields has been criticised for degrading the land and requiring the liberal use of pesticide and fungicide. Furthermore, although land is dramatically better used for bananas than beef in terms of nutrition per hectare, there are still parts of the world in which forests are being cleared for banana plantations[8] (see Deforestation, page 154).

Overall, however, the only really bad bananas are any that you let rot in your fruit bowl. These join the scandalous 30 per cent of food wasted by consumers in the UK and many other countries. If you

---

* I bought a bunch in a plastic bag weighing just 4 g, the reason for which is to stop customers ruining them when they try to split a bunch. So the bag could be worth it until we all learn not to maul the fruit.

do find yourself with bananas on the turn, they are good in cakes and smoothies. I have a distant childhood memory that they are also tasty in custard.

An orange

90 g CO$_2$e each or 500 g per kilo average
1 kg CO$_2$e each or 5.5 kg per kilo air-freighted for the start of a season

Most oranges, along with most apples and bananas, are great from a carbon perspective.[9] They keep well, and so can be grown in natural conditions and shipped around the world to wherever they are required.

The important thing to note here is that although there are lots of food miles, these ones are fairly climate friendly. Like bananas, oranges can go on a huge boat and take their time. However, I was told by someone who buys fruit commercially that some supermarkets airfreight some varieties of orange at the start of the season to get them into the shops a couple of weeks early. I estimate that a litre of orange juice has a footprint equivalent to around 6 kilos of oranges – many more than it would take to produce that much juice. That's because orange juice incurs several inefficiencies in its production:

- The pulp is thrown out (so 'with bits' varieties and smoothies may be more sustainable).

- There are emissions from processing, including pasteurising and sometimes turning into concentrate for transport purposes, and refrigeration.

- There is the footprint of the carton.

- Transport miles are often higher as the product moves from farm to juicer to cartoner to distributor, sometimes zigzagging wildly around the world.

- Fresh orange juice requires refrigeration. Tesco report that their freshly squeezed juice has about twice the footprint of the long-life product. Most of that difference will be down to refrigeration.

An hour's TV

34 g CO$_2$e 15-inch LCD flat screen
76 g CO$_2$e 28-inch CRT TV
88 g CO$_2$e 32-inch LCD flat screen
220 g CO$_2$e 42-inch plasma screen

One hour per day on the 32-inch LCD comes to 32 kg CO$_2$e per year – equivalent to a 45-mile drive in an average petrol car.

Overall, watching TV turns out to be a remarkably low-carbon hobby, and beats anything that involves driving. This is good news because the average European spends a massive three and a half hours a day in front of the box.[10] You probably don't because you read books and, surely, there isn't time for both.

At its very worst, the 42-inch plasma screen, on for 10 hours per day, could clock up 800 kg CO$_2$e per year,[11] the equivalent of driving an average petrol car for about 1100 miles. That may sound like a lot, but it actually makes for quite a low-carbon life, because it leaves so little time in your day to do anything else that might have a higher footprint.

The figures above don't take account of the emissions embodied in the TV set itself. The significance of these emissions – relative to the power the TV actually consumes in use – depends on what TV you have and how often you use it. The figure of 220 kg of CO$_2$e is a ballpark figure for the manufacture of a brand new TV costing £500, which at the time of writing is about the price of the energy-hungry 42-inch plasma version. This works out as 22 kg per year if you keep it for 10 years. If you watched that TV for 1 hour per day, the emissions from the electricity it will use, at about 80 kg per year, will still dwarf those of the manufacture. At the other end of the scale, if you spend £200 on a 15-inch LCD, make it last just 5 years and watch it for only half an hour a day, the embodied emissions will dominate your TV footprint.

By watching with friends you can clearly make things more efficient. The more people you invite around the better, provided they live within walking or cycling distance.

Should you replace your TV?

At my local waste disposal centre (that's the place that used to be called the tip in the days before segregation) they currently have a whole room especially for homeless old-fashioned CRT televisions, most of which work fine but which are being disposed of to make way for modern flat-screen models. The people who run the disposal centre say that, at the peak, they were taking in 400 CRT TVs per week.

So what are the carbon implications of trading in an old television for a new one? Figure 3.1 provides some answers. All the sums are based on these assumptions: that your old TV is a typical 28-inch CRT model; that whatever choice you make now, you will stick with it for 10 years; and that you will watch 1 hour of TV per day throughout that time.

In short, my sums indicate that sticking with your old TV is a good idea unless you're happy to switch to something smaller. There are two clear winning options, each with a similar viewing experience and costing about the same over the 10-year period: a new energy-efficient 15-inch flat screen or a second-hand 14-inch CRT. Although the 15-inch flat screen has the lowest energy use, the 14-inch CRT wins overall at just 35 g per hour including the satellite receiver. But if you keep your TV for longer than 10 years the winning option on every count is to buy the 15-inch LCD.

If you don't want to switch to a small screen, however, sticking with the 28-inch CRT screen is the best option because the embodied energy of its manufacture has already been written off.

So the message is that although getting a new TV does give most people a chance to improve their energy efficiency, if you don't buy carefully, it is likely to do the reverse.

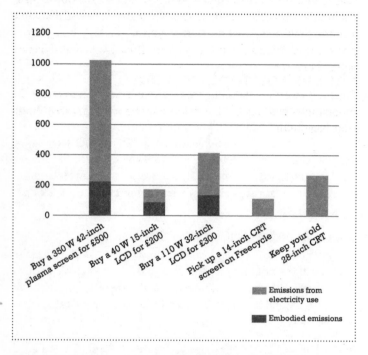

Figure 3.1. The carbon footprint of different TV options, based on watching for 1 hour per day and not replacing again for 10 years.

What about standby?

TVs typically use about 3 watts in standby mode, but since that probably accounts for at least 20 hours out of every 24 it means that your TV could well cause 15 kg of emissions over the course of a year even when there's nothing showing on the screen. If you have a small, efficient TV, that could be the biggest part of its annual footprint. Only you can decide whether standby adds enough quality to your life to justify the 15 kg. My recommendation is that you cut it if you can but don't let the issue torment you. If you spend a lot of time in front of the box, the additional exercise of switching off by hand will probably raise your quality of life slightly.

Lots of different devices around your home, all on standby at once, could collectively be more significant and it should be said that some standby circuits use a lot more power than 3 watts. With a plug-in power meter costing about £10 you can check. No house should be without one.

Finally, it's worth mentioning that it also takes carbon to create the programmes you watch – but that is a whole new story (see World Cup, page 160).

100 grams
to 1 kilo

A mug of tea or coffee

21 g CO$_2$e black tea or coffee, boiling only what you need
53 g CO$_2$e with milk, boiling only what you need
71 g CO$_2$e average, with milk, boiling double the water
you need
235 g CO$_2$e a large cappuccino
340 g CO$_2$e a large latte

**So if you drink four mugs of tea with milk per day,
boiling just what you need, that's the same as a
60-mile drive per year in an average car. A single
latte every day would be nearly 1 per cent of the
10-tonne lifestyle.**

The shock here is the milk. If you take tea or coffee the British way,
with milk, and you boil only the water you need, then the milk
accounts for two-thirds of the total footprint (see Milk, page 71). The
obvious way to slash the footprint of your tea is reduce the amount
of milk, or simply to take it black (herbal tea, anyone?) (Figure 4.1).
Although this will reduce your nutritional intake, you could easily
replace the lost calories with something more carbon-friendly such
as a biscuit.

I have based my cappuccino and latte sums on the large kind that
some of the coffee-house chains encourage you to quaff. These come

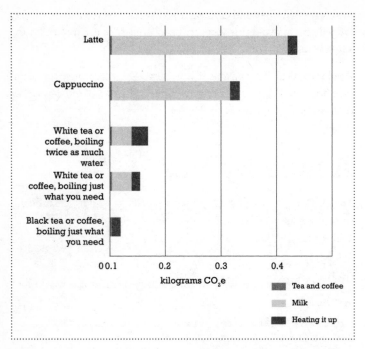

Figure 4.1. The footprint of a 250 ml mug of tea or coffee with no sugar.

in with a higher impact than four or five carefully made Americanos, filter coffees or teas. They also mean you are drinking an extra half a pint of milk, perhaps without realising it.

At my work we've suddenly decided that next week we're all going to do without milk in our drinks. At worst it will taste horrible. At best we'll change habits of a lifetime, resulting in decades of reduced hassle, lower carbon, slight cost savings and possibly even fractionally improved health. It has to be worth trying.*

If you boil more water than you need (as most people do), you could easily add 20 g CO_2e to your drink. Boiling more than you need

---

* Stop Press. Update: we survived. It was horrible. I'm going to pick different battles. A little bit more herbal tea is drunk in the office these days, possibly as a result of the experiment.

wastes time, money and carbon; if you haven't yet developed perfect judgement, to avoid this you can simply measure the water into the kettle by using a mug.

Finally, think about your mugs. Buy sturdy ones; look after them and save hot water by only washing them up at the end of the day, rather than using a fresh mug for every cup.

A mile by bus

15 g CO$_2$e one of 20 passengers squeezed into a minibus in the suburbs of La Paz
150 g CO$_2$e typical London bus passenger
1.4 kg CO$_2$e per mile Lake District resident sharing a double-decker bus with just the driver

The efficiency of a bus is just about proportional to people it is carrying. It also depends on the amount of stopping and starting.

A double-decker bus can do between 7 and 14 miles to the gallon. A couple of years ago I took one of these through the Lake District from Windermere to Keswick. It was just me and the driver for the whole ride. At, say, 10 passenger miles per gallon it worked out far worse than leaving the bus in the garage and me taking the car. Another way to look at it is that the bus was going anyway, and my getting on it was just about carbon free. It's a Catch-22. No one wants to take the bus, because it is just as cheap and quicker to take the car (if you've got one), and so far people who care about the carbon arguments are still in a pitiful minority. On the other hand, if the bus were three times as frequent and one-third of the price it would probably be very popular. Everyone apart from the car manufacturers would be better off.

La Paz, Bolivia, is the place I think of where this principle is practised to perfection provided you are prepared to set aside a bit of safety and comfort. Twelve-seater minibuses charge around town with 20 or more people crammed inside. You can get just about anywhere for one Boliviano – a few pence – and you are unlucky if you have to wait more than 5 minutes. Most people in the developed world would choose a luxury version of this for perhaps five times the price, but the principle is sound and in Bolivia 10 years ago the 'value proposition' met the market need perfectly.

All my numbers have factored in the fuel supply chains as well as the exhaust pipe emissions. I also include a component for the emissions entailed in manufacturing the vehicle, although for the bus this is a small consideration because they do so many miles before needing replacement.[1]

A nappy

89 g CO_2e reusable, line-dried, washed at 60°C in a large load, passed on to a second child
145 g CO_2e disposable
280 g CO_2e reusable, tumble-dried and washed at 90°C

So that's 550 kg per child for two and a half years in disposables; the equivalent of nearly two and a half thousand large cappuccinos.

Most parents will be relieved to hear that there is usually no carbon advantage to be had from reusable nappies. On average they come out slightly worse, at 570 kg per child compared with 550 kg for disposables. And if you wash them very hot and tumble-dry them, reusables can be the worst option of all. However, if you put your mind to it you can make reusables the lowest-carbon option. To do this, pass them on from child to child (so that the emissions embedded in the cotton are spread out more), wash them at a lower temperature (60°C), hang them out to dry on the line, and wash them in large loads.

For a disposable nappy, most of the footprint comes from its production. But about 15 per cent arises from the methane emitted as its contents rot down in landfill (contrary to the myth that if you wrap them up in a plastic bag they will never rot at all).

The study I'm basing my figures on assumed that the average child stays in nappies for about two and a half years, and is changed just over four times a day.[2] On this basis, in the UK, nappies account for something like one two-thousandth of total greenhouse gas emissions – or more like half a per cent for homes with babies.

What does all this mean for the carbon-conscious family? If you have two children and stick to non-tumble-dried reusables throughout, you might be able to save nearly half a tonne CO_2e. You will also cut out landfill. It's a significant efficiency, but (here's the catch) you need to know your own minds before you start out because if you give up, revert to disposables and bin the reusables, it could be the option with the highest footprint of all. But try to keep all of this in perspective: if you take just one family holiday by plane you will undo the carbon savings of perfect nappy practice many times over.

UK climate change secretary Ed Miliband recently drew on the same nappy study to defend his announcement that his own children wear disposables. He was roasted – somewhat unfairly I thought – by blogging eco-mums who claimed that the study was fatally flawed. Poor chap. At least he'd thought about it. The debate illustrates, yet again, that this kind of analysis is more murky and subjective than we might think.

A punnet of strawberries

150 g CO_2e (or 600 g per kilo) grown in season in your own country
1.8 kg CO_2e (or 7.2 kg per kilo) grown out of season and flown in, or grown locally in a hothouse

How have we got into the habit of buying tasteless out-of-season strawberries, which have a footprint more than 10 times the tastier seasonal version?

Although I've given just one number for local, seasonal strawberries, the precise footprint depends on such things as the soil, the use of fertiliser and the use of polytunnels.[3] Some of these variables increase both the yield and the emissions per hectare, so whether they result in more or less carbon per strawberry is not so simple to work out. Luckily, they are all so much better than the out-of-season version that a good enough rule of thumb is just to stick to those grown in your own country – unless your government subsidises the heating of greenhouses (as is the case, for example, in The Netherlands). This kind of hot-housing is, broadly speaking, just as bad as air-freighting the fruit from hotter countries (see Flying, page 135, and Asparagus, page 83).

In short, then, the best advice is to wait until they are in season, then enjoy them twice as much. Or if you really can't wait, buy frozen or tinned: these lie somewhere in the middle of the range, in carbon terms, along with those travelling 'middle distances' by road and boat from warmer climes.

All the figures here have taken account of the 23 per cent average wastage between the field and the checkout. A small amount of the footprint is the packaging and this is actually in a good cause if it enables more of the strawberries to find their way into our mouths. The footprint of the plastic will typically be lower than that of the wasted fruit.

A mile by train

0.15 kg CO_2e Intercity standard class
0.16 kg CO_2e London Underground
0.19 kg CO_2e light rail or tram
0.30 kg CO_2e Intercity first class
An 18-mile intercity rail journey has the same footprint as a cheeseburger, whereas a 5-mile journey on the Tube is equivalent to a pint of milk

Although trains can be a relatively green way to get around, the figures above show that the emissions of rail journeys are higher than you might think. All the numbers provided include the direct emissions and electricity consumption of the moving train itself but also attempt to take account of the embodied emissions from train manufacture, the upkeep of the rail network and the running of all the infrastructure.[4]

The amount of energy required to propel a train down a track depends mainly on just a few simple things:[5]

- How fast the train goes. The air resistance goes up with the square of the speed.

- How many stops there are. Each stop wastes energy – the exact amount being proportional to the square of the speed and the weight of the train. Some newer trains reduce this stoppage waste through 'regenerative braking', a similar technology to the one used in hybrid cars.

- Rolling resistance of the wheels on the track. This is lower for trains than for cars because metal wheels on metal tracks are more efficient than rubber tyres on asphalt. The rolling resistance goes up proportionally with the weight of the train.

- The type of fuel used. Electricity beats diesel because although there are inefficiencies in generating it from fossil fuels in the first place, once this has been done the train engine can turn almost all of the power into movement. A diesel engine is much less efficient.

Long-distance Intercity trains go fast (that's bad) but stop infrequently (that's good). In the UK, they're often electric (that's good), but they're also extremely heavy (that's bad). The weight of the train per passenger seat, amazingly, is around twice that of an average car. Just to be clear, what I am saying is that the weight of a full train is twice that of all the cars that would be needed if each passenger drove instead. Professor Roger Kemp,[6] who has looked at this astonishing fact in detail, explains it in terms of overengineered safety: trains weigh at least twice what they need to because we have become obsessed with safety and have forgotten that rail travel is already over

100 times safer than driving. A couple of miles from my house an Intercity train derailed and rolled down a high embankment. Incredibly, only one person was killed. The event was still splashed across the national news, raising public fears, even though so many more people die on the roads every single day. One price of this excessive focus on safety may well be that twice as much energy is required to get our trains moving every time they leave a station.

First-class travel deserves a mention because the number of seats you can squeeze into a first-class carriage is around half the number in a standard-class carriage. This means that the weight being moved per person is doubled again; we're now up to the weight of four cars per seat. I sometimes board trains where half the length is nearly empty first class and the rest is crowded standard class, suggesting that the real weight being hauled per first-class passenger may be even higher.

Things are a bit more complex when it comes to the Eurostar, because when it's in France it runs on electricity that comes predominantly from nuclear power. This is low-carbon energy, whether or not you think nuclear power is worth it in other ways. However, I don't think it is useful to think of trains in nuclear-friendly France as having a smaller footprint than those elsewhere – which is how they are sometimes portrayed. That's because all the nuclear electricity that French power stations can produce would get used up regardless of whether any trains were running. In that sense, the trains are effectively powered by the fossil fuel plants that provide the extra electricity over and above the nuclear 'baseload' (see A unit of electricity, page 55, for more on this somewhat confusing concept of marginal depend).

Interestingly, the London Underground is almost as low-carbon, per passenger mile, as Intercity trains, despite stopping much more often. This is mainly because people are packed in so tightly – almost tessellating, nose to armpit. Other reasons are that the Tube travels relatively slowly, is all-electric, and has lighter trains.

Overall, trains are generally a lot greener than cars but not as good as walking, cycling or staying at home. A sensibly designed car can win, provided you fill it with people. Even two people travelling together are better off driving an efficient car than travelling first class.

See also London to Glasgow return, page 117.

A 500 ml bottle of water

110 g CO$_2$e locally sourced and using local distribution
160 g CO$_2$e average
215 g CO$_2$e travelling 600 miles by road

A bottle a day would add up to 0.6 per cent of the 10-tonne lifestyle.

At more than 1000 times more carbon intensive than its tap alternative, knocking bottled water out of our lives has got to be a simple win. It doesn't even taste better.

Processing the water is the easy part: the bulk of the emissions come from packaging and transport. There is 80 g CO$_2$e per litre just for the plastic. On top of that is the energy required to melt the PET (polyethylene terephthalate) balls down and mould them into bottles. Transport is significant because water is so heavy. If it has gone 600 miles by road, that could add a further 115 g CO$_2$e per bottle.[7]

As I write this, London has announced plans to start reintroducing public drinking fountains. This is an encouraging step forward. If everyone switches away from bottles it will be great for the environment and still just as healthy, refreshing and convenient. Interestingly, even though people will be financially better off, the economy may look as though it has slowed down a fraction. This is a nice illustration of how inadequate it is to measure how we are doing by our economic growth. When we are all using the fountains, we might collectively look a shade poorer on paper because the few people who make their living persuading us to buy the bottled stuff will need new jobs. But that will be more than compensated for by the extra cash that the average person will save. So the economy will recede as we all get better off. Let's not cry for the peddlers of bottled water either. Even if you don't believe that they had it coming to them, they are clearly talented and persuasive people who are also more than capable of being successful in constructive careers.

If the world consumes 200 billion litres of this bottled water per year,[8] that's 80 million tonnes of greenhouse gases, or one-sixth of a per cent of global emissions. This is a win worth having!

A letter

140 g CO_2e a 10 g letter made from recycled paper and recycled by you
200 g CO_2e a typical 25 g letter printed on virgin paper and sent to landfill
1600 g CO_2e a small catalogue sent to landfill

If you have five letters delivered per day plus two catalogues per week, that's a massive 480 kg CO_2e per year, nearly 5 per cent of the 10-tonne lifestyle.

Mail clocks up a carbon footprint in four basic ways (Figure 4.2):

■ Paper production. The carbon footprint of paper manufacture depends on the recycled content, the quality of the paper and the efficiency of the mill. The junk mail coming through our door generally uses high-quality stuff and doesn't tend to boast any recycled credentials. My estimates are based on paper that has a typical UK mix, with less than one-fifth recycled content. That gives it a footprint of 2.35 kg CO_2e per kilo. The best estimate for pure virgin paper comes in at 2.59 kg per kilo, and 100 per cent recycled paper at about half of that; it takes about half as much energy to create new paper from old paper as it does to create paper from trees.[9]

■ Printing. For the footprint of printing on the paper to turn it

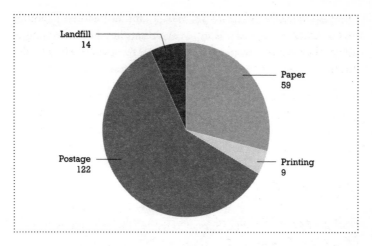

Figure 4.2. The carbon footprint of a 25 g letter, printed on virgin paper, posted by second-class mail and thrown into landfill (grams CO_2e).

into glossy and enticing sales literature, I estimate an additional 350 g CO_2e per kilo.

- Postage. For a standard letter, this accounts for most of the footprint. It's impossibly difficult to trace the carbon footprint of posting a letter by direct means. However, if you take the footprint of the postal services sector as a whole and divide it by the turnover of that sector you can get a broad idea of the carbon footprint per unit of cost. In the UK it comes out at about 380 g CO_2e per £1 spent. A 25 g second-class letter would have cost 32p in the UK, and we can associate a carbon footprint of about 120 g CO_2e with that. So most of the impact of a junk letter comes from the burden that it places on the whole infrastructure of our postal system: vans, trains and sorting offices.

- Decomposition. A good deal of junk mail ends up in landfill, where it decomposes anaerobically and produces methane. For this I have allowed 550 g CO_2e per kilo of paper.[10] You can avoid this, of course, by recycling as much mail as possible. This is OK to do even if the letter has a plastic window. But do remove any other plastic – such as film wrap.

Eliminating junk mail will declutter your life as well as saving carbon. The purpose of most of it is to persuade you to buy stuff you don't need, so brain purification is probably the biggest reason of all for putting an end to it. To avoid junk mail, use a free junk-mail opt-out service. In the UK you can write to the Mailing Preference Service, Freepost Lon20771, London WE1 0ZT. Give your address and a list of all the names of people to be taken off all possible mailing lists.

The service can take a few months to kick in, but it should work. At the back of this book, just to make things easy for you, there is a page ready for you to cut out and send.

To deal with the ones that still get through, keep a stack of printed labels by your door saying 'Return to sender. Please strike us off your database.'

In the UK, there's a Royal Mail service for avoiding unaddressed junk mail that is delivered to everyone on postal rounds. If you wish to opt out of receiving Door to Door mail items, email your name and address to **optout@royalmail.com**. Note that there are caveats on the Royal Mail website about the other unaddressed mail that this will stop.

Finally, a message to the instigators of junk mail: more and more people will think badly of you for using high-carbon marketing techniques. If you must use mailshots, at least keep your databases clean, use recycled paper and keep your messages short.

Sending an email beats sending a letter hands down (see page 15).

1 kg of carrots

0.25 kg CO_2e local, in season
0.3 kg CO_2e average
1 kg CO_2e shipped baby carrots

So a bag of carrots is like a 2-mile train ride.

At around 2 g CO_2e per calorie, these and other root vegetables are some of the most climate-friendly foods available – and healthy too. If you ate only these foods and others that have similar carbon intensity you could feed yourself for just over 1 kg CO_2e per day, or less than 500 kg CO_2e per year.

Seasonal vegetables have small carbon footprints because they avoid all of the main greenhouse gas sources for food: they are grown in natural conditions without artificial heat, they don't go on aeroplanes, and they don't incur the inefficiencies inherent in the production of food from animals.

If you go on to boil your carrots for 10 minutes, you will add a few more grams CO_2e per kilo to the footprint. (For more on cooking, see boiled potatoes, page 69.) My children will only eat their carrots raw. That suits me fine. It's better from every angle – there's less carbon emission, it saves time, and the nutritional value is better.

Note that some baby varieties have a much lower yield per acre of land, resulting in higher emissions per kilogram. So it usually makes sense to buy full-sized, classic varieties. And, as with other vegetables, favouring misshapen specimens may help avoid wastage in the supply chain (see page 183).

A newspaper

0.3 kg CO_2e the *Guardian Weekly*, recycled
0.39 kg CO_2e the *Sun*, recycled
0.48 kg CO_2e the *Daily Mail*, recycled
0.82 kg CO_2e the *Guardian*, recycled
1.8 kg CO_2e a weekend 'quality' paper, recycled
4.1 kg CO_2e a weekend 'quality' paper, sent to landfill

A quality paper every day of the week adds up to 270 kg CO_2e per year, even if you recycle them all. That's equivalent to flying from London to Madrid one way.

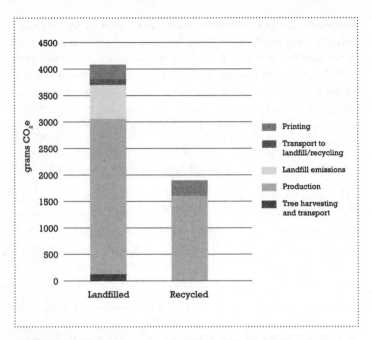

Figure 4.3. The carbon footprint of a weekend newspaper. Sending paper to landfill causes methane emissions *and* means that more carbon-intensive virgin paper has to be produced.

It's amazing how energy-hungry newspaper production can be. And the figures provided here are on the low side, because none of them take account of the footprint of journalism itself – including the newspaper offices and staff flights. At the highest end of the spectrum, just a Sunday paper each week could add up to 2 per cent of a 10-tonne lifestyle if you don't recycle it (Figure 4.3). At 1.25 kg, the weight of a weekend paper is part of the problem. In our house only a tiny fraction would be read, so the rest might as well never have been printed.

The reasons why recycling is so important are twofold. First, if paper is disposed of in landfill sites, it emits methane as it rots. Second, for each newspaper that isn't recycled, one more newspaper's worth of virgin paper has to be manufactured. For these reasons, throwing your paper in the general waste more than doubles its footprint.[11]

Opting for a slimmed-down weekly paper, such as the *Guardian Weekly*, is one good way to reduce emissions. Another is to get your news online. If you do this for an hour a week on a 50-watt laptop and if we multiply that by, say, 5 to take account of the production of the laptop, the running of your network and the electricity consumed by all the hubs and servers around the world that support the websites you browse, it still comes to around half the impact of the *Guardian Weekly*. If only I could take my laptop into the bath ....

A pint of beer

300 g CO$_2$e locally brewed cask ale at the pub
500 g CO$_2$e local bottled beer from the shop or a pint of foreign beer in a pub
900 g CO$_2$e bottled beer from the shop, extensively transported

A pint of local ale per day in the pub would be 1 per cent of the 10-tonne lifestyle. A few bottles of imported lager per day might be as much as 10 per cent.

The beer at the low end of the scale is based on figures for the Keswick Brewing Company, a microbrewery quite near where I live. Just about everything you can think of was included in the study I did for them (Figure 4.4). There were the obvious things such as ingredients, packaging, fuel, electricity and transport. I also included such elements as staff travel, the carbon cost of having to replace their equipment every so many years, and office stationery.

For the Keswick Brewing Company, I estimated that ingredients accounted for about one-third of the footprint, fuel and electricity about another one-quarter, and staff travel about one-tenth. The fermentation process itself releases CO$_2$, accounting for about one-twentieth (15 g per pint). Most of the company's beer is sold in reusable casks, so the footprint of packaging is kept right down.

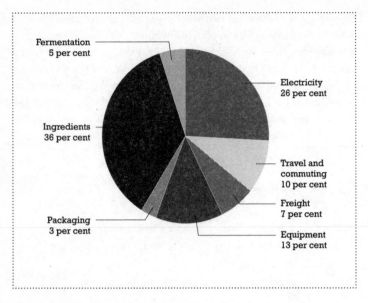

Figure 4.4. The footprint of cask beer from the Keswick Brewing Company.

Distribution is 7 per cent, even though all their deliveries are fairly local, because beer is such heavy stuff.

A few miles from the Keswick Brewery is another, larger brewery. Delivery from there to pubs just down the road is via a distribution centre in Wolverhampton, a couple of hundred miles away. This is the usual story for big breweries and their subsidiaries. Even the country of origin is not always obvious from the branding. Although a few hundred road miles are not usually the most significant factor for foods, beer is an exception because it's so heavy. Hence opting for local ale is usually a good idea.

For home consumption, and thinking for a moment only of carbon rather than taste, cans are slightly better than bottles, provided you recycle them. (I can feel the connoisseurs at Keswick cringing as I write.) Heeding this advice is especially important if the beer is travelling a long way because the glass also adds to the weight.

Wherever and whatever you drink, a single pint of a quality beer is almost always better for both you and the world than spending the same money on several tins of bargain-basement brew.

Finally, as though it were needed, the carbon angle gives yet another reason not to drink and drive (see Car crash, page 141).

A bowl of porridge

82 g CO_2e traditional Scottish, made with water
300 g CO_2e about half milk (just how I like it)
550 g CO_2e milky and sweet

So a bowl of traditional Scottish porridge is equivalent to a 90-second mobile-to-mobile phone call.

	Footprint (grams CO_2e)	Calories	Calories per kg CO_2e
Milky and sweet			
Oats (50 g)	40	185	
Full-cream milk (350 ml)	448	315	
Bringing to the boil, stirring carefully	42	0	
Sugar sprinkled on top (10 g)	20	40	
Total	550	540	981
Traditional Scottish			
Oats (50 g	40	185	
Water (350 ml)	0	0	
Bringing to the boil, stirring carefully	42	0	
No sugar allowed	0	0	
Total	82	185	2256

Table 4.1. The carbon footprint of porridge

A bowl of half-milk porridge every day would be about 1 per cent of the 10-tonne lifestyle. Cement and porridge made like this don't just look the same, they also have very similar carbon intensity per kilo.

Oats is a fantastic low-carbon food that also happens to be healthy and tasty. If you fed yourself entirely on milky sweet porridge it would cause just 900 kg CO_2e per year. By sticking to the Scottish water-based version you'd cause a trifling 340 kg CO_2e per year: about one-tenth of the typical UK diet. Table 4.1 shows how the footprint breaks down. As with a cup of tea (see page 35), it's the milk that dominates.

The cooking is about half of the footprint of the traditional Scottish version. I've assumed that you cook it on the stove and never have the lid on because you are stirring like crazy, trying to save yourself a washing-up nightmare. A non-stick pan should help. Better still, the microwave is lower carbon than an electric hob or gas ring and doesn't cause sticking; but keep a close watch or it will turn into an exploding mess from Doctor Who. Enough said about all this. I am the last person who should be writing a cookery book.

A shower

90 g CO_2e 3 minutes, efficient gas boiler, aerated showerhead
500 g CO_2e 6 minutes in a typical electric shower
1.7 kg CO_2e 15 minutes in an 11-kilowatt electric power shower

If you have high-carbon shower habits, there could be half a tonne per year to be saved here – equivalent to a return flight from London to Madrid.

At the low end of the spectrum, 3 minutes is how long I take if I wake up half an hour before my train is due to leave. Gas is a more efficient way of providing heat than electricity, provided you have a reasonably efficient boiler. The aerated shower head helps by making less water feel like more. In theory at least, it saves water and carbon without you having to forgo any comfort at all.

If you are in a family of four and you each spend 15 minutes in an electric shower every day, you may be able to reduce your household footprint by a tonne per year just by switching to an aerated shower-head. Switch to a gas-powered shower and there's another half tonne to be saved. Finally, you can cut the remaining emissions by a factor of three by having 5-minute showers – and it is only at this point that you are having any impact on your lifestyle. You will be swapping time in the shower for time doing almost anything else that you want: reading a book, lying in bed, both of these at once, or whatever you like. If you take all these measures your family could knock off 2 tonnes per year – and save about £350, which would easily pay for the couple of paperbacks you might each luxuriously read in bed during the time you have liberated.[12]

The showers in Iceland are worth a mention as the most luxurious I've ever had. Geothermally heated and almost zero CO_2e, they are all the more enjoyable after a day out in the abundant rain and snow there. Unfortunately you have to fly to get there. (See also Bath, page 81.)

An ice cream

50 g CO_2e a 60 g ice lolly from the supermarket, eaten on the day of purchase
500 g CO_2e a big dairy ice cream from a van

The ice lolly is essentially frozen sugary water, and in the supermarket the refrigeration is likely to be relatively efficient.

At the high-carbon end the dairy ice cream's footprint is higher for three reasons: it's bigger, it's dairy based and it's been kept cold in a much less efficient mobile refrigeration unit. The inclusion of dairy ingredients means that all the inefficiency of ruminant livestock farming has been incurred. The van fridge is much less efficient than the one in the supermarket.

My figures are just based on cigarette-packet calculations. I've guesstimated from a broad understanding of the footprint of different food ingredients and transport impacts and from knowing a little bit about mobile refrigeration.

A unit of heat

50 g CO_2e using a solar water-heating panel
244 g CO_2e using a modern (90 per cent efficient) gas boiler[13]
400 g CO_2e using an old, 55 per cent efficient gas boiler
600 g CO_2e from UK grid electricity
1060 g CO_2e from Australian grid electricity

By a 'unit' I mean 1 kilowatt-hour. That is enough to run a 'one-bar' electric fire for 1 hour, or enough to boil about 15 litres of water in an electric kettle.

At the low end, the solar water heating panel has no operational emissions. I haven't given it a zero CO_2e rating because the manufacture of the panel itself will have a carbon footprint. The exact number depends on factors such as the design of the device, where it is used and how it is maintained, so my figure is really just a guesstimate. One problem with solar heating is that it tends to be 'low-grade' heat. In other words it's all right for warming up baths and gentle heating of rooms, but not usually much good for boiling kettles or making toast.

In the middle of my scale is heat generated by an efficient gas-fired boiler, such as might power a new central heating system. In this scenario your heating is done by fossil fuels but at least you're using them fairly efficiently: the only losses will typically be around 10 per cent of the energy disappearing out of the flue.

(That said, in the case of a central heating system there may also be inefficiencies caused by heating rooms that you are not actually using. If the only room you want to heat up is the kitchen, the most efficient thing you can do may be to turn on the gas cooker. That way nothing goes up the flue and all the heat goes into the room you want to keep warm.)

At the high end of the scale is electricity. This is a 'high-grade' form of energy that can be used for many different things, so it's generally a waste to use it just for heating. The precise footprint depends on which country you are in (see page 197), but with only a few exceptions the figure will always be high because the electricity is usually generated from fossil fuels and – unlike with a gas boiler in your home – more than half the energy in the fuels is lost in the power station or transmission grid. In other words, it's generally inefficient to use electricity just for heating. In the UK, the average unit has a footprint of about 600 g CO_2e,[14] whereas the figure is higher, for example, in coal-dependent Australia.

A unit of electricity

60 g CO_2e from the Icelandic grid
600 g CO_2e from the UK grid
900 g CO_2e from the Chinese grid
1060 g CO_2e from the Australian grid

The carbon impact of using an additional unit of electricity is often higher than we're led to believe.

Electricity generation is one of the principal causes of carbon emissions all over the world. However, as we've seen, the exact carbon cost of each unit of power depends on the precise mix of generating fuels used in your country. Icelandic electricity comes almost exclusively

Country	Direct emissions from power generation, per unit consumed	Estimate of total footprint per unit consumed (including the carbon cost of extracting fuel from the ground, maintaining and building power stations, wind turbines, etc)	Estimated footprint of marginal demand: e.g. the carbon cost or saving of each unit of electricity you choose to use or save
Australia	1.00	1.06	1.06
China	0.84	0.90	1.06
Iceland	0.00	0.06	0.06
Norway	0.00	0.06	1.06
UK	0.54	0.60	1.06
U.S.	0.60	0.66	1.06

Table 4.1. The carbon footprint of electricity consumption in different countries. The marginal demand column shows that unless you live in Iceland, someone somewhere is likely to have to burn more coal if you use more electricity.

from fossil-fuel-free geothermal and hydropower plants, so the only footprint comes from creating and maintaining the infrastructure.

Australian and Chinese electricity, by contrast, comes mainly from highly polluting coal. UK electricity is somewhere in the middle, coming from a mixture of coal, gas (which is less polluting than coal but is still a fossil fuel), nuclear (which has a low carbon footprint but is contentious in other ways) and a smattering of renewables.

Most people who think about carbon footprints are used to the idea that each unit we consume causes a fixed quantity of CO_2 emissions. However, the truth is somewhat more complex than that. A more meaningful way to think about the carbon footprint of your own electricity use is to think of it as being additional to all the power consumption that was already going on before you flicked on the light or appliance. Looked at this way, the extra demand that you place on the grid is met entirely through additional fossil fuels, because the renewables in your country will already be running at full capacity. In other words, when you turn the lights on, you don't personally affect the amount generated by renewables because they are already going flat out. Rather, what you trigger is almost certain to be a lump of coal thrown into a power station. This is the case throughout Europe, because even in countries where all electricity comes from renewables or nuclear, adding to demand reduces the amount of electricity that those countries are able to export, thereby increasing the fossil-fuel generation in other nations. Looked at in this sense – in terms of 'marginal demand' (see Table 4.2) – each unit of electricity you consume has a footprint of at least 1 kg CO_2e per unit, regardless of which country you live in.

One exception is Iceland, where, for the moment at least, it does look as though you can more or less use as much electricity as you like without boosting your footprint. The country is overflowing with hydroelectric and geothermal power: you can *see* the energy almost everywhere you go, boiling out of the mud and pouring over water-falls. But once Iceland works out how to export its clean energy, or how to import enough of the world's heavy industry to use up the renewables capacity, electricity will become a scarce resource for Ice-landers, just as it is for the rest of us. In the meantime, enjoy it!

The great green tariff swindle?

'Green' or 'renewable' electricity tariffs and suppliers may sound attractive, but the hard reality is that signing up is unlikely to reduce the climate change impact of your electricity signifi-cantly. This applies whatever the colour of the company's logo or however ecological the company name might sound.[15]

The two main claims made by the 'green' providers are that elec-tricity comes from renewable sources and/or that they use the money you spend on your bills to invest in a new renewable capacity. Neither of these is necessarily what it might appear.

The 'from renewable sources' claim

All electricity suppliers in the UK are obliged to submit Renewable Obligation Certificates (ROCs) to the government for up to 10.4 per cent of the electricity they sell to their customers. They can get these certificates either from generating their own renewable power or by buying them from others. Suppose a company has a tariff in which all electricity is sourced from renewables. It sounds great. However, this means that the supplier gets a lot more ROCs than they need to hand over to the government. The normal practice is for the 'green supplier' to sell these to other suppliers, thereby allowing them to simply source less of their own power from renewables. So the net carbon benefit is zero – but the 'green supplier' is quids-in because it has managed to charge you a premium. The tariff only makes a difference to the extent that the provider retires some ROCs (tears them up) instead of selling them on. In the UK, Good Energy claim to do this with 5 per cent of them, although this has been challenged, with some people suggesting they have only been retiring 2 or 3 per cent. It doesn't much matter, because they are arguing over such low percentages. The main point is that well over 90 per cent of the ROCs are kept in circulation. If you switch to the 'green tariff' offered by one of the larger electricity suppliers, the chances are that no ROCs are retired at all and you are allowing them to worsen the energy mix in their other tariffs, while using the 'green' story line as a way of charging you a premium.

The 'investing in renewables' claim

What if a company claims that it will invest so much of every pound you spend via the company in new windfarms and other renewable energy projects? This sounds great but what it could boil down to is that the supplier is simply engaging in two different business activities. One is being an electricity provider just like all the others, but with 'green' branding. The other part of the enterprise is investing in renewable power generation. Both of these could potentially be good business opportunities regardless of any environmental considerations. The key question is whether the investments in the new windfarms would still be made if you got your electricity from elsewhere. Is the company promising to invest to the tune of your electricity spend in projects that would otherwise not go ahead at all? In other words, is it genuinely *additional*? This is a very long way from being clear to me. One thing that is certain is that it is possible to run a roaring commercial enterprise along these lines.

I am not saying that the companies claiming to provide greener electricity aren't greener than average. They probably are, and I do get my own electricity from one of them. What I am saying is that their impact may not be quite as low as you think. The overall message is that if you want to reduce the footprint of the electricity you buy via the grid, the only real way to do it is to consume less.

Spending £1

Minus **330 kg CO_2e** on a well-executed rainforest preservation project
Minus **3 kg CO_2e** on solar panels
160 g CO_2e on financial, legal or professional advice
720 g CO_2e on a car
930 g CO_2e on a typical supermarket trolley of food[16]
1.7 kg CO_2e on petrol for your car
4.6 kg CO_2e on flights[17]
6 kg CO_2e on the electricity bill
10 kg CO_2e and beyond on budget flights

Unless you are deliberately investing in something that reduces emissions elsewhere, it is just about impossible to spend money without increasing your carbon footprint. Everything causes ripples of economic activity and with it emissions. So with wealth comes carbon responsibility. I'm hardly the first person to have suggested this but it's an important concept. So what are you going to do with your £1?

If all your money goes on travel you may be at the worst end of the irresponsibly wealthy. If you invest it in forests and wind farms you are at the opposite extreme, using your wealth to bring about a low-carbon world. If you spend a million pounds on fine art, you are mainly passing on the responsibility for doing the right thing with that cash to the artist or the dealer. If you stick it under the mattress, it is doing neither harm nor good.

Of the specific examples given above, flying gets to be such a high-impact way of spending cash for two reasons. First, the aviation industry can buy its fuel for around 30p per litre. Second, it then burns it at an altitude where it has, as a best estimate, nearly twice the climate change impact that it would have had at ground level. So spending money on jet engine fuel is six times as carbon intensive as spending the same money on petrol and putting a match to it.

Leaving the lights on is another of the cheapest ways of trashing the planet, suggesting that for all the talk of higher fuel prices we are a long way from establishing a serious financial incentive to go green.

My petrol figure is based on a true cost of 40p per mile for an average petrol car. This takes account of the extraction, shipping and refining of the fuel, but not the depreciation or maintenance of the car.[18]

At the positive end, I have included some of the fairly limited options for actually doing carbon friendly things with money. They range enormously in their effectiveness, which is something not all policy makers seem to have fully grasped.

On page 200 is a larger list of the carbon intensity of different industries whose goods and services you might buy. No industry sector in our model comes in below 160 g per pound sterling. The more you think about this, the clearer it becomes that there is simply no avoiding the advantages of slowing the economy down or of changing its structure. We could do with spending less time charging around earning as much as we can to buy things we don't really need. We would do well to become better at enjoying what we've got – and to disentangle our self-esteems from our pay packets. Without wishing to sound like a sandal-wearer, it's clear that we've become locked into a mindset that is not going to serve us well over the coming decades. If you're not convinced, have a read of Tim Jackson's book *Prosperity Without Growth*.[19]

1 kg of rubbish

200 g CO$_2$e garden waste
700 g CO$_2$e average bin contents
9 kg CO$_2$e aluminium and copper

The average UK person sends 330 kg to landfill or incineration[20] each year and recycles just 170 kg. This causes 230 kg CO$_2$e, which is 2.3 per cent of a 10-tonne lifestyle.

By 'rubbish' I mean things you dispose of by putting them in the normal bin as opposed to recycling or composting it. Looked at this way, there are two parts to the footprint.

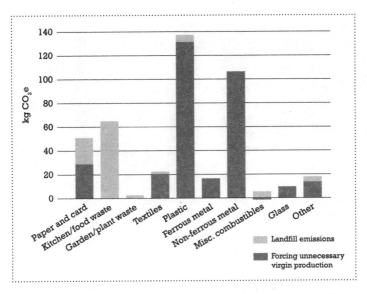

Figure 4.5. The annual footprint of the average UK householder sending waste to landfill rather than to recycling or composting.[21]

First, there are the landfill emissions, which are due mainly to stuff rotting down underground, without air. This anaerobic decomposition produces methane, only some of which gets captured and the rest escapes to warm the world. (This isn't an issue for metals, glass and building materials, of course, because they don't rot down in the way that food, paper and garden waste do.) There is also a little bit of fossil fuel required to run a landfill site.

Second, there is the fact that by not recycling something you are forcing more virgin materials to be produced for use in future products. This isn't an issue for food, for which recycling was never an option. But for metals, textiles, plastics and paper it is a big deal.

Figure 4.5 shows that recycling our aluminium and plastic is where most of us can make the biggest improvements. That's mainly because it takes so much more energy to make a brand-new aluminium can or plastic bottle than it does to make a new one from an

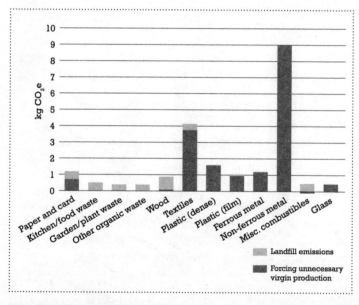

Figure 4.6. The footprint per kilo of throwing stuff into landfill compared with recycling or composting it. In other words, this graph shows the *difference* that recycling makes.

old one. Kitchen waste is a key area, too, because of the large amount of methane it produces when it rots underground.

Figure 4.6 shows that when you are standing with 1 kg of something in your hand, if that something is aluminium, it is particularly important that you recycle it. The next most important per kilo are textiles.

The significance of food waste is underplayed in both these graphs because they don't take account of the footprint of needless production. These graphs just show the difference between landfill and recycling or composting.

Washing up

Almost zero CO_2e (but the plates aren't clean) by hand in cold water
540 g CO_2e by hand, using water sparingly and not too hot
770 g CO_2e in a dishwasher at 55°C
990 g CO_2e in a dishwasher at 65°C
8 kg CO_2e by hand, with extravagant use of water

Running a dishwasher twice a week on the economy setting comes to 80 kg per year, equivalent to a 110-mile drive in an average car.

The results of the great dishwasher versus handwashing debate are as follows. The most careful hot-water handwashing just about beats the dishwasher but loses out badly on hygiene (nearly 400 times the bacteria count on the dishes) and time (four times as long as loading the dishwasher). Overall the dishwasher wins, particularly because the figures here probably don't reflect the most energy-efficient machines that are now on the market. I also haven't included the carbon saving that is possible if you set your machine to run in the middle of the night when electricity demand is low and the grid becomes more efficient.

The handwashing figures are based on a study of people around Europe,[22] but I've used the UK electricity mix to calculate the carbon. (If you live in nuclear powered France, don't be fooled into thinking your electricity consumption doesn't matter so much. It all gets traded around, as discussed on page 55).

My figures for the dishwasher are based on always running a full load and they include 130 g CO_2e for the wear and tear on the machine itself (based on a fairly expensive, 'built to last' model that you keep for 10 years[23]). The conclusion, then, is get a dishwasher. It simultaneously helps the planet, your health and your lifestyle. When you buy one, choose a make that will last, and look after it. Try to always run it full, use the economy setting when possible and run it in the middle of the night if you can because the electricity is less carbon intensive.

I haven't included anything for the detergent or the mains water consumption, because they are nothing compared with the impact of heating the water.[24]

A final note: I have known people routinely wash their stuff by hand before putting it in the dishwasher. This must be the worst of all options and ranks alongside ironing your spouse's socks for needless slavery (see Ironing, page 22) If this is your routine, please consider yourself liberated.

A toilet roll

450 g CO_2e recycled paper
730 g CO_2e virgin paper

If you have typical North American wiping habits, that comes out at 75 kg CO_2e per year or three-quarters of a per cent of the 10-tonne lifestyle.

The typical North American supposedly uses 57 sheets of toilet roll per day. That seems excessive to me, although I haven't been counting. The figure comes from the **ToiletPaperWorld.com** website and surely they must know these things.[25] The Worldwatch Institute puts annual consumption at 23 kg per year for a North American, 1.8 kg per year for an Asian and just 400 g for the average African.[26]

I'm not sure I want to launch into a detailed exploration of bathroom technique here, but because three-quarters of a per cent of the 10-tonne life seems high for such a simple and brief part of our lives it does seem worthy of a moment's personal reflection. My numbers show that a sense of economy is in order. If, as I suspect, many of us could halve our usage without any negative side-effects, then it's an easy and worthwhile carbon win.

I'm not advocating hardship here, just calling for a simple perspective check; are our backsides in their rightful place or are they getting spoilt? Have decades of adverts talked us into believing that a pampered bum is one of the hallmarks of a rich and fulfilled life? My footprint figures here are based on numbers from Tesco, whose research suggests a carbon cost of 1.1 g per sheet for their recycled stuff and 1.8 g for traditional paper. So that's three spam emails for a sheet of recycled,[27] five for virgin or two sheets of virgin for one genuine email.

Driving 1 mile

344 g CO_2e a Citroen C1 doing a steady 60 miles per hour
710 CO_2e an average car achieving a UK-typical 33 miles per gallon
2240 g CO_2e a Landrover Discovery, new but not looked after, doing 90 miles per hour

So driving a car the average annual distance of 9000 miles would use between 3 and 20 per cent of the 10-tonne lifestyle, depending on the type of car and how you drive it.

At the low end of the scale we have four people travelling together in a well-maintained low-emission vehicle (such as a Citroen C1, Peugeot

107 or Toyota Aygo) travelling at a steady 60 miles per hour. My own C1 can do 65 miles to the gallon under these conditions. With four people, the carbon comes out at 86 g CO_2e per person mile.

At the high end of the scale we have a single person in a poorly maintained, rapidly depreciating, high-emissions car that looks more like a tank, cruising at 90 miles per hour or driving unsympathetically in urban conditions with heavy use of both brakes and accelerator. In these conditions, a vehicle of this type may achieve as little as 11 miles per gallon.

My numbers are higher than those you normally see for driving. That is partly because I am including the emissions from the extraction, refining and transportation of fuel as well as just the burning of it. Even more importantly, I am factoring in the manufacture and maintenance of the vehicle itself.

As a rule of thumb, about half of the carbon impact of car travel comes out of the exhaust pipe itself.[28] A few per cent come from the processes of extracting, shipping, refining and distributing the fuel (see Petrol, page 87). The rest, typically 40 per cent of the footprint, is associated with the manufacture and maintenance of the car. Big, expensive new cars have more of their embodied emissions attributable to each mile of driving. An older car that is still fairly efficient could beat my C1 by virtue of having had its embodied footprint written off. (See New car, page 143.)

But it's not just what model you drive that matters. Here are 10 good ways to reduce the carbon footprint of your car use:

- Use the train, bus or bike if travelling alone. Typical saving: 40–98 per cent. (See London to Glasgow return, page 117.)

- Put more people in the car. This could make it better than train travel provided that the others were otherwise going to drive separately. Typical saving: 50–80 per cent.

- Join a car sharing scheme.

- Drive a small, efficient car. Typical saving: 50 per cent compared with the average car.

- Look after your car so that it will do 200,000 miles in its lifetime and runs as efficiently as it can. Typical saving: 30 per cent compared with the average. (See A new car, page 143.)

- Accelerate and decelerate gently, avoiding braking where possible. Typical saving: up to 20 per cent in urban conditions.

- Drive at 60 miles per hour on motorways. Typical saving: 10 per cent compared with 70 miles per hour.

- Keep the windows up when driving fast, and the air-conditioning off. Typical saving: 2 per cent.

- Keep the tyres at the right pressure. Typical saving: 1 per cent.[29]

- Avoid rush hour. (See Congested car commute, page 107.)

- Drive safely. (See Car crash, page 141.)

Is it worth slowing down?

Although we might know that driving more slowly on motorways is better for the planet, this concern often gets outweighed by our desire to get there on time. After all, time is money, right?

I'm going to take the case of someone driving on their own and assume that they value their time at time at £16 per hour. That's about the take-home pay per hour of someone earning £40,000 for a 37.5-hour working week plus a half-hour motorway journey (around 35 miles) to the office each way. This is above average for the population as a whole but may be about typical for those who commute on motorways.

I have assumed that the commuter in question drives a car that is capable of 40 miles per gallon at 70 miles per hour but 54 miles per gallon at 60 miles per hour. That is reasonable because, in motorway conditions, the fuel consumption is roughly proportional to the square of the speed.[30] I'm also going to assume that this person hardly cares at all about their impact on climate change (we're going to look at financial costs and benefits only) and that the carbon cost of petrol is 3 kg CO_2e per litre (see page 87).

As Table 4.3 shows, in this scenario, the slower driver saves carbon without losing money. Those less well off or with hungrier cars would be better off as well.

	70 mph	60 mph
Value of driver's time, per hour	£16	£16
Miles per gallon	40 mpg	54 mpg
Carbon footprint of a 70 mile round-trip commute	21 kg CO_2e	16 kg CO_2e
Time cost of the 70 mile commute	£16	£19
Cost of petrol	£8	£5
Total cost	£24	£24

Table 4.3. How to save carbon without losing money

A red rose

Zero CO_2e picked from your garden, no inorganic fertiliser used
350 g CO_2e grown in Kenya and flown by air
2.1 kg CO_2e grown in a heated greenhouse in the Netherlands

A single red rose could have the same impact on climate change as about four and a half kilos of bananas.

Could the banana ever replace the rose on Valentine's Day? If you try this low-carbon alternative, please let me know how you get on.

The numbers here sum up the Hobson's choice that you are faced with if you want out-of-season cut flowers. You either have to put them on a plane or grow them using artificial heat. Both of these are bad news for climate change.

The study I based my numbers on found that for consumers in the UK, Dutch roses had about six times the carbon footprint of the air-freighted ones.[31] After all, Holland is a cold country in winter and roses take a long time to grow. This only adds up commercially because The Netherlands subsidises the energy required by its flowers industry. In the UK, home-grown flowers will probably have enjoyed only the sun's heat.

In my work on UK supermarket products, out-of-season cut flowers emerged as some of the products with the largest carbon footprint per pound generated at the tills. In other words they are one of the most carbon-unfriendly ways of getting rid of your cash.

There's another concern, too. All commercial cut flowers use land that could otherwise be growing food. The demand for agricultural land is already driving deforestation (see page 154), which in turn is responsible for around 18 per cent of man-made emissions. Looked at in those terms, cut flowers have to mean less rainforest – so the true footprint is probably even bigger than my numbers suggest.

Quite a few people I've spoken to have said that their attraction to cut flowers has wilted once they have made the connection with the huge emissions and pressure on land that they bring about.

So, stick to your own grown garden crop if you can, and do without flowers when they are not in season. As for alternatives, longer-life indoor plants are a dramatically less carbon-intensive option. And some artificial flowers are just about indistinguishable from the real thing – if you can bear the concept.

1 kg of boiled potatoes

620 g CO$_2$e locally grown, boiled gently with the lid on
1170 g CO$_2$e still local but boiled furiously with the lid off

This panful of potatoes contains two-thirds of a woman's daily calorific needs. If potatoes were all you ate for a year you could feed yourself for just 330 kg CO_2e, or 3 per cent of the 10-tonne lifestyle. That is good going when you consider that food and cooking currently accounts for 3 tonnes CO_2e per person per year. (That's without taking account of deforestation, which could add half as much again). You'd end up bored and malnourished if you stuck rigidly to this regime, of course, but there is clearly a place for potatoes in the low-carbon lifestyle. Table 4.4 shows how the footprint breaks down.

	Grams CO_2e
Growing the potatoes	220
Transport	80
Packaging in a simple bag	10
Supermarket storage and display	60
Boiling	250 to 800
Total	**620 to 1170**

Table 4.4. Breaking down the potato footprint

Potatoes are a low-carbon crop; larger conventional varieties are especially so, simply because yields are higher.

Transport emissions are not high, provided these local potatoes stay in the locality. It is not uncommon for some supermarkets to move produce hundreds of miles to a distribution centre and then back again. Even when this happens, however, the transport does not have a disastrous impact.

The biggest part of the footprint comes from the cooking process. The way you do this can alter the total footprint by a factor of two. Here are some ways to keep the cooking emissions to a minimum:

- Use a gas cooker.

- Use a lid on the pan.

- Boil gently. The temperature of the water, and therefore the cooking speed, is exactly the same when you turn the gas down to a gentle simmer as when you boil at full throttle.

- Cut the potatoes into smaller pieces.

- Use a pressure cooker: the pressure raises the boiling temperature, which means the potatoes cook faster and more efficiently.

Alternatively, if you are baking or roasting, you can do the following:

- Use a microwave or a fan-assisted gas oven.

- Reduce the size of pieces.

- Having heated the oven up, cook more than one thing.

I have ignored the carbon cost of getting to the shops (see Driving a mile, page 65, and Cycling a mile, page 23).

A pint of milk

723 g CO$_2$e

So if you get through two pints a day in your household, that's 527 kg per year, as much as a return flight from London to Madrid.

Milk is high-carbon stuff for exactly the same reasons that beef is. Cows, like most animals, waste a lot of the energy in the food they eat in the process of simply keeping warm and walking around rather than creating meat and milk. In addition, cows ruminate (chew the cud) which means they burp up methane, roughly doubling the footprint of the food they produce.[32]

As Figure 4.7 shows, around 85 per cent of the milk's footprint takes place on the farm, but transport, packaging and refrigeration also play their part. Because milk is heavy, keeping it local (and not trucking it hundreds of miles to and from distribution centres) seems like a good idea. My instinct is that milkmen probably cut carbon footprints by keeping the weight of our shopping bags down and therefore making it that much easier to walk to the shops for everything else. In addition, reusable glass bottles almost certainly beat plastic disposables, even if you recycle the latter every time.

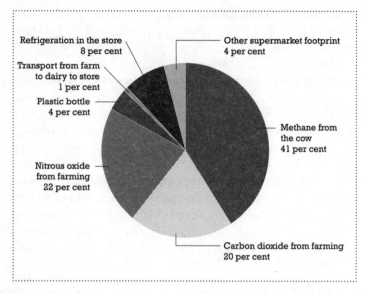

Refrigeration in the store
8 per cent

Transport from farm
to dairy to store
1 per cent

Plastic bottle
4 per cent

Nitrous oxide
from farming
22 per cent

Other supermarket footprint
4 per cent

Methane from
the cow
41 per cent

Carbon dioxide from farming
20 per cent

Figure 4.7. The carbon footprint of locally sourced milk in a plastic bottle at the checkout of Booths Supermarkets. In this example, the milk comes from Bowland Fresh, a local supplier, so the transport impact is low. This chart doesn't include your journey to the shops, or home refrigeration.

Wherever you get your milk, however, it remains – like all food from cattle – a high-carbon way to get your calories. There is probably quite a lot that could be done to reduce its carbon cost but it's a hugely complicated area to research. Various studies have been carried out so far but they don't always agree. To give a flavour of how confusing everything is, if you change the feed, you alter the carbon cost of that feed, the milk yield *and* the amount of methane that gets belched out. At the same time you play about with factors like the life expectancy of the cow, the amount of saleable meat that the herd will produce alongside its milk and the other inputs that will be required to keep the cow healthy. To make things even more complex, different farming practices affect the ability of the soil to absorb and store carbon. And everything also depends on the location of the farm and the breed of cow. Nobody has yet properly worked out how all these variables interact.

If the carbon footprint were the only consideration, the unpleasant truth is that the most efficient thing to do would probably be to keep cattle in small indoor spaces and rear them as intensively as possible, minimising wasteful activities such as getting exercise or keeping warm. But carbon isn't the only consideration, of course, especially for organic farmers such as David Finlay in southwest Scotland. David is reducing the milk output of his herd in southwest Scotland from 7500 litres per cow per year to 5000. He believes that although the milk yield will go down, the amount of meat he can sell will go up and his feed costs will fall, along with his use of antibiotics and other inputs that have both financial and practical costs. Hugely important to David are two other factors: the animals will have even better lives than they already have on his farm, and he stands to have more free time because he will only have to milk them once a day. He believes a system like this could compete in the supermarkets alongside conventionally farmed milk even without the organic label. In fact, he believes the price premium on the organic label comes mainly from the administrative costs of demonstrating at every step of the journey from farm to shop that no contamination with conventional milk has taken place.

One proposed solution to the belching problem is Rumencin, a simple additive that markedly cuts methane production in cows. The EU classified it as an antibiotic and banned it, even though the farmers I've spoken to say this was a mistake because it does not have the human health impacts usually associated with antibiotics. I'm not an expert on these things, but I can believe this might possibly have been a bureaucratic blunder that is now waiting to be overturned.

Whatever the truth about different dairy farming practices, soya milk is almost certainly a lower-carbon option than anything from a cow. Even though I haven't seen a study of this, in comparison with cows' milk there is none of the inefficiency of putting animals in the food chain and no rumination involved. The market for soya is driving deforestation, but the problem is not the stuff that is eaten directly by humans: most soya is fed to ... cows.

1 kg of cement

100 g CO$_2$e Eco-Cement
710 g CO$_2$e standard cement, efficient production
910 g CO$_2$e global average
1 kg CO$_2$e inefficient production

The world produces around 2.2 billion tonnes of cement per year – or around 300 kg per person. Nearly half of this (47 per cent) is produced in China. Making this basic building material results in a staggering amount of CO$_2$e: around 4 per cent of the world's total greenhouse gas footprint.[33] This figure is so high because the chemical process that turns limestone into cement gives off large volumes of CO$_2$ directly *and* takes a huge amount of energy.

Around half the footprint is down to the chemical reaction. There is not much you can do to reduce this without changing the product itself. About 40 per cent comes from the burning of fuel to drive the reaction, leaving 10 per cent for other bits and bobs in the cement industry and its supply chains.

Because of the basic chemical reaction required to make the stuff, it is hard to see how conventional Portland cement could be made into a low-carbon product. One alternative is Eco-Cement, a product invented by John Harrisson in Tasmania. Eco-Cement's advocates claim not only that this product requires half the energy input of conventional cement, but also that it reabsorbs CO$_2$ from the air as it hardens (around 400 g CO$_2$e per kilo). There are also claims that it is easier to incorporate waste materials into the mix than with normal cement and that it is easier to recycle. The product is based on magnesite, which is not as abundant as limestone, and perhaps that's why not everyone is using it yet. Or perhaps it is no good at sticking things together. I haven't tried it.

Cement makes up about 12 per cent of the footprint of the UK construction industry, so other potential ways of reducing its impact are to use different materials, to build to last and build less, and to refurbish in preference to knocking down and building anew (see House, page 149).

1 kilo *to* 10 kilos

A paperback book

400 g CO$_2$e recycled paper, with every copy printed getting sold
1 kg CO$_2$e average
2 kg CO$_2$e the same book on thick virgin paper, with half the copies getting pulped

The carbon footprint of a typical paperback is about the same as watching 12 hours of programmes on an average TV.

Overall, reading is a low-carbon activity and there is plenty of room for it in the sustainable lifestyle.* Why? It's hard to drive or shop while you read. For a short while, a gripping novel halts the consumerist lifestyle in its tracks.

My average figure is based on a 250 g book printed on paper from a UK-typical mix of virgin and recycled pulp.[1] I've assumed that 60 per cent of all copies made are actually sold, even though I've heard more pessimistic estimates than this. The economies of scale in printing

---

* 'Sustainable lifestyle': this is a tricky expression. It doesn't bear much scrutiny and we could get hopelessly bogged down defining it. However, I strongly suspect that whatever your definition I would still stand by my assertion that it leaves plenty of scope for reading.

are such that it pays to print too many.

At the high end, the same book is printed on heavyweight high-gloss virgin paper and weighs 350 g. Half of the print run is pulped without ever hitting the shops.

At the low end, the book still weighs 250 g but is printed entirely on recycled paper. Roughly speaking, it takes about twice as much energy to make paper from trees as it does from recycled pulp – though the actual value varies enormously depending on the efficiency of the paper mill and the quality of the paper.

What you are reading right now doesn't yet exist as I write, but I'm guessing that, in carbon terms at least, you are holding a better-than-average paperback because my publisher thinks about these things. However, once you stop to think about it there are all sorts of difficult questions about what to include in the sums. I haven't included the electricity burned by my computer as I'm typing right now, or any part of the footprint of my publisher's offices at Profile, or a host of other possible elements.

Nonetheless, I hope this book pays for itself in carbon terms fairly easily. You only have to cut out about three car miles to cancel out its production.

All carbon footprints need to be thought of in terms of 'bang for buck': do the benefits outweigh the impact? To maximise the 'bang' side of the equation you simply have to read this book, talk about it and pass it around.

Electronic book readers deserve a mention. I guesstimate that an e-reader has a footprint of around 50 kg.[2] If I'm right you'd have to get through at least a hundred paperbacks (bought new and then sent to recycling) before the paper saving outweighed the embodied emissions of the reader itself. This is before electricity consumption of the reader and in IT networks has been taken into account. E-readers may be wonderful devices but I can't see a carbon argument for getting one, unless it gets you reading more. And you can't yet take them in the bath, either.

A loaf of bread

1 kg CO_2e an 800 g loaf

Bread is good stuff: a year's calorific intake can be had for around half a tonne CO_2e. That's only 5 per cent of the 10-tonne lifestyle and one-sixth of the current UK diet.

As Figure 5.1 shows, just over half the emissions of a loaf of bread come from the actual growing of the ingredients. About one-sixth is the baking. Transport is typically one-seventh, and the supermarket itself adds about one-ninth. The bag is a very small consideration – and if it helps to keep the bread fresh for longer, it is probably well worth it.

Bread is a great low-carbon food provided we actually eat it. There's the catch. It gets thrown away because we are fussy eaters and because it doesn't keep well. Tristan Stuart's eye-opening book *Waste* has a picture of a Marks & Spencer sandwich factory systematically dis-

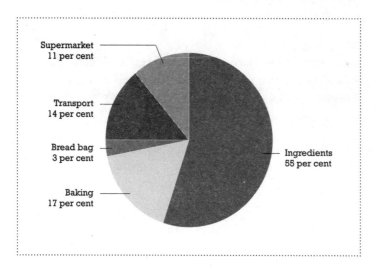

Figure 5.1. The footprint of bread at the supermarket checkout.

carding four slices from every loaf: the crust and the next slice from each end.[3] The remaining slices get made into fresh sandwiches and are still at risk of being binned before they are sold. Only once safely through the checkout do the odds of a sandwich being eaten start looking good, but there are still such hurdles as children who won't eat crusts and over-catered corporate lunches.

Loaves sold straight to consumers are no better, because the shelf life is so low. Plenty is binned by the supermarkets and plenty more goes stale in bread bins, or ends up in a half-eaten sandwich. To keep the carbon cost of your bread to a minimum, buy only what you need, enjoy the crusts and get your children to do the same. Find uses for stale bread: toast, dunked in soup and so on. Remember that bread mould doesn't kill you. And buy smaller loaves if you are not getting through them – the introduction of the 600 g loaf will help with this.

A bottle of wine

400 g CO_2e from a carton, with few road miles
1040 g CO_2e average
1500 g CO_2e over-elaborate bottles, transported for thousands of miles by road

So if you drink three bottles of typical wine per week, which is pushing the limits of a healthy lifestyle, that is about 150 kg per year, equivalent to driving 210 miles in an average car.

My estimates here are based on a study I did for Booths supermarkets (Figure 5.2). For a typical bottle, just over one-third of the footprint comes from the production of the wine itself. Whether or not it is possible to reduce this by buying organic wine is not clear, although there may be other environmental benefits of the organic option. It is difficult to know from the label what the carbon intensity of a particular vineyard is, so I have just given all wine a typical value, based on various studies.

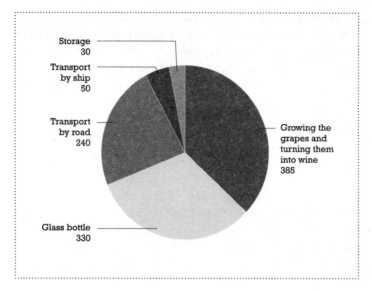

Figure 5.2. The carbon footprint of a bottle of wine (in grams CO_2e).

The glass bottle accounts for a similar amount of carbon to that of the wine it contains. There is a simple saving to be made here: by buying wine boxes or cartons you can reduce the footprint of the packaging by a factor of about five. In doing so you will also reduce the weight, so transport emissions can also be slashed by one-third. There will be absolutely no loss of quality, even though you might lose some choice. If the carton offends you, you can always decant the wine into a jug.

There is a lot that can be done without getting rid of the glass altogether. Organico is a wine distributor near where I live that has started importing some of its wine unbottled. This cuts the transport weight. It does its own corking and puts a £2 deposit on the bottles, which are themselves 15 per cent lighter than normal and are made from clear glass because this is better for eventual recycling. One further nice touch is that they have done away with the concave bit under the bottle that has always struck me as fundamentally dishonest.

Note that shipping is only a small component, so it doesn't matter all

that much what continent your wine comes from. Far more important are the road miles – both in your country and in the country of
origin. For this reason, locally produced wine could cut the footprint
by 25 per cent, provided that your neighbourhood has the right kind
of climate.

Because it is less dilute, wine often turns out to be a slightly less carbon-
intensive way of taking alcohol on board than beer (see page 49).

All these calculations assume that you recycle any packaging.

1 kg of plastic

0.75 kg CO_2e EcoSheet
1.7 kg CO_2e PET for plastic bottles, from recycled
materials
3.4 kg CO_2e polystyrene from virgin materials
3.5 kg CO_2e average
4.4 kg CO_2e polypropylene for injection moulding, made
from virgin materials
9.1 kg CO_2e some types of nylon[4]

Plastic is such useful stuff: it's tough, durable and waterproof. No
wonder we use so much of it. Unfortunately, plastic tends to be so
durable that it hangs around in landfill sites for centuries, clutters
up the stomachs of animals and fish, transforms remote Scottish
beaches into junkyards and has ended up in almost every ecosystem
you can think of. But from a purely carbon perspective, its inability
to rot is good news in as much as it won't add to methane emissions
from landfill: if we assume that the plastic is put in the bin rather
than tossed into a street or field, those hydrocarbons are going back
underground where they came from.

As the figures above show, the footprint of making plastic from virgin
material is about double what it would be if recycled products were
being used. The challenge for recycling plastics is that it's difficult but
necessary to separate the various types and process them separately.

This isn't true for EcoSheet, however. This brand-new construction material can be made from the full range of different plastics, so almost nothing goes to landfill. Once you have finished with it, the sheeting can even be reworked into new boards. The makers, 2K Manufacturing, told me that they don't even need to heat up waste plastic to the usual recycling temperatures to create their boards. As I type, only a few sample boards per day are being produced, but by the time you read this, full production is expected to have been underway[5] and there is every likelihood that the Science Museum in London will have used the stuff to build its new exhibition on climate change.

Biodegradable plastic packaging is worth a mention because it can be a well-intentioned disaster area. It sounds great, but if you send it to landfill it rots down and emits methane and if you throw it into the recycling bin it can ruin the entire batch. It should be compostable instead, but I have also heard that it releases chemicals that slow down the degrading process for the rest of the bin or heap.

Taking a bath

Zero CO_2e heated by solar energy
0.5 kg CO_2e modestly filled, efficient gas boiler
1.1 kg CO_2e generously filled, efficient gas boiler[6]
2.6 kg CO_2e generously filled, electric water heating

A daily bath adds up to between 180 and 950 kg CO_2e per year – that's between 2 and 10 per cent of the 10-tonne lifestyle.

In our family at least three of us often end up using the same water, even if not all at the same time. (Anyone who's been running through mud has to go last.) Since we top up with hot, the bath is always full to the overflow by the end. That is about 120 litres, giving a footprint per person of around 400 g.

If you were to read a book in the bath for an hour, you'd probably add 50 per cent to the footprint of the average full bath by pulling out the plug with your toes from time to time and topping up with hot. So the actual leisure activity would be 500 g per hour on top of the functional bath itself. That is quite a bit higher than most TV watching but still a lot lower than any pastime that involves using a car.

In winter you can reclaim about half the heat simply by leaving the plug in until it goes cold. This works provided that you actually want the heat in your bathroom and don't object to the idea of old bathwater hanging around.

The comparison with showers (see page 52) might be a surprise. Electric showers on the market range from 7 kilowatts at the weedy end to 11 kilowatts at the powerful end. For the same impact as a full bath from an efficient gas boiler you could have a 9-minute power shower or a 13-minute weedy shower. In comparison with sharing bathwater you would have to be a family of fast scrubbers to make the electric shower win out. That's even without taking account of any bathwater heat reclaimed (which isn't an option for the shower unless you have a plug and are prepared to stand in an ever-deepening pool). To be fair, though, showers where the hot water comes from a gas boiler, which is increasingly the norm, are much more carbon-efficient and generally will work out as using less energy than a bath unless the latter is shared between many people.

What about other options? A trip to the swimming baths could have a much higher impact than a bath, even if you were to walk there (see page 152), whereas a wild swim comes out best of all – though not many of us live near a clean and safe river or lake.

Overall, baths do serve a purpose and even the most luxurious needn't be too bad as long as they are not electrically heated. On the other hand, if everyone in your household has extravagant bathing habits it could easily come to over 1 tonne per year.

A pack of asparagus

125 g CO$_2$e a 250 g pack, local and seasonal
3.5 kg CO$_2$e the same pack, air-freighted from Peru to the UK in January

If your entire diet were as carbon intensive as long-haul asparagus, your food would have a footprint of more than 50 tonnes: three and a half times the entire annual footprint of an average UK citizen.

The numbers here are based on data from Booths supermarkets, who to their credit took steps to increase their UK sourcing when they saw the impact of the Peruvian product and are now emphasising the benefits of seasonal food more strongly than ever. Air-freighted from Peru, asparagus comes in at 14 kg CO$_2$e per kilo or, to put it another way, 70 g of carbon per calorie. It is over 100 times more carbon-efficient to get your calories from bread.

When 1 kg of produce is being moved, a mile by air has more than 100 times the climate impact of a mile by sea. This is because it takes a lot of energy to keep a plane in the air – and also because engine emissions tend to do more damage at high altitude than they do at ground level (see Flying from London to Hong Kong, page 135). For this reason it is difficult to see how there can be any place at all for air-freighted food in a sustainable world.

Examples of other foods that are very likely, when out of season, to have been air-freighted or (just as bad) grown in an artificially heated greenhouse include baby corn, baby carrots, mangetout, bobby beans[7] (green beans), fine beans, okra, shelled peas, lettuces, blueberries, raspberries and strawberries.

At the other end of the scale is asparagus grown in season in your own country. This cuts out a staggering 97 per cent of the footprint. When asparagus is out of season (which is most of the time), try to favour low-carbon options such as kale, carrots, parsnips, swedes or leeks.

Flying from closer-by north Africa has considerably less impact than flying from Peru. And at each end of the local asparagus season there are periods in which a small amount of heating makes the crop viable.

None of the estimates here include the footprint of cooking the food, which is likely to be around 100 g CO_2e if you simmer it for 8 minutes with the lid off.

A final comment: the recipe book I consulted advised strongly against air freight on taste grounds, stressing the importance of eating asparagus within 48 hours of harvesting.

A load of laundry

0.6 kg CO_2e washed at 30°C, dried on the line
0.7 kg CO_2e washed at 40°C, dried on the line
2.4 kg CO_2e washed at 40°C, tumble-dried in a vented drier
3.3 kg CO_2e washed at 60°C, dried in a combined washer–drier

If you wash and dry a load every two days that's equivalent to 440 kg CO_2e, which is like flying from London to Glasgow and back with 15-mile taxi rides to and from the airports.

Modern washing powders work just as well at 30°C, so there is a very simple saving to be had here of 100 g CO_2e per wash just by turning the temperature down. But the bigger savings relate to drying. As the numbers above show, for a typical 40°C wash nearly three-quarters of the carbon footprint comes from the drying rather than the washing. Tumble driers generally use electricity to generate heat. This is more than twice as carbon intensive as generating heat from gas. If you use a conventional vented drier, most of the heat is simply pumped out to the outside world, so overall it's a wasteful activity. Condensing driers use a little bit more energy still although, in winter at least, all that heat stays inside your house, where it is probably of some use.

Overall, a household running the tumble drier 200 times a year could save nearly half a tonne by installing a clothes rack inside and a washing line outside. In winter the evaporation from drying clothes will cool your house down slightly, but it's a marginal effect and on a baking hot summer's day our clothes drying in the kitchen act as free air-conditioning.

Make sure that your washer has a good spin function. It is much more efficient to remove most of the water by spinning it off than by evaporating it.

All the figures listed above are based on a full 5 kg load (half loads use a little less energy each time but they work out as much less efficient per garment washed). I've allowed around 220 g per wash for the embodied emissions in the appliances themselves.[8] If this estimate is correct, the manufacture and delivery of the appliances account for nearly 10 per cent of the total carbon footprint of each wash.

You can probably improve on the lifetime of your washer and/or drier if you look after it and get it repaired when it breaks. Switching from a typical 1998 machine to a new one with an 'A' rating might gain you around 10 per cent in efficiency;[9] in other words, roughly enough to offset the emissions required to make the new machine but no more. So the message is that unless your machine is particularly cranky and inefficient there is no real carbon case for getting a new one unless you have to.

While on the subject of washing, have you optimised the frequency with which you wash stuff? I don't want you to start going around smelly, but it's worth asking the question: does stuff go in the wash unnecessarily often? If you can reduce the number of washes you do without upsetting anyone, there is a time saving to be had, too, so it's a great example of life getting better as the carbon comes down.

A burger

1 kg CO_2e veggieburger
2.5 kg CO_2e 4-ounce cheeseburger.

If you eat a cheeseburger each day that's a massive 910 kg CO_2e per year – the same as driving 1500 miles in a fairly efficient car and just over 1 month's worth of ration in the 10-tonne lifestyle.

The 4-ounce cheeseburger described here provides 515 calories. If this were the only type of food you ate, the average man would need about five burgers per day provided he didn't take much exercise. (The average woman could get away with one fewer.)[10] If you managed to keep up this diet for a year without killing yourself, you'd cause about 4.6 tonnes of carbon emissions just through your food.

The cheeseburger's footprint breaks down as shown in Table 5.1.

Component	Grams CO_2e
Beef (108 g)	1910
Cheese (20 g)	250
Bun (40 g)	50
Salad (20 g)	10
Condiments (20 g)	80
Cooking and transport	200 (approx.)
Total	2500

Table 5.1. The carbon footprint of a 4-ounce cheeseburger

The biggest factors here are the beef and the cheese. As we've already seen, animal produce tends to be more carbon intensive than vegetables and grains because animals consume a lot of energy just to keep themselves warm and move around. This makes their job of converting animal feed into meat and milk inherently inefficient.

There is another big problem with beef and dairy farming. Cows, like sheep, are ruminants. This means that they belch out methane, a greenhouse gas 23 times more potent than CO_2. The result is that beef and lamb have around double the carbon footprint per kilogram of meat compared with that from pigs.

A further consideration is that excessive demand for meat provides an incentive for deforestation because it raises the demand for arable and grazing lands. That said, there is plenty of land, for example in the UK, that is fit for cattle and sheep farming but not for crops, and there can be a conservation benefit in having animals on the land.

It is unclear whether the footprint of the burger could be reduced by using organic or free-range meat and cheese. The Cranfield University study[11] I have used for my figures found that organic cattle farming had little or sometimes negative carbon benefits. However, the organic farmers I know who have studied this report were scathing about the assumptions made about organic practices and yields. The carbon benefits of rough grazing are also unclear. On the one hand less feed is required, but on the other hand there are complex implications on yield and rumination.

At the time of writing my inclination is to say that a low-carbon diet can safely contain a bit of beef and lamb from rough pastures, but the whole area clearly warrants further research. And in the meantime, there's always the low-carbon veggieburger option.

See also Steak (page 95) and Deforestation (page 154).

A litre of petrol

3.15 kg CO_2e

In the UK we get through about 50 billion litres of petrol and diesel per year. That's something like four bathfuls, right up to the overflow, for every man, woman and child in the land.

The pie chart in Figure 5.3 speaks for itself. If you were to pour a litre of petrol on the floor and strike a match, the fumes would account for only about three-quarters of the carbon story. The other quarter is caused by the supply chain of the fuel: getting it out of the ground, flaring off the gas, shipping it around the world, refining it and getting it to the pump.

This extra quarter doesn't usually feature in car emissions statistics (including official greenhouse gas 'conversion factors'), which generally deal only with the stuff that comes out of the exhaust pipe. This is one part of the reason why the carbon footprint of driving is often so badly underestimated.

The story for diesel is slightly different. Each litre has a slightly higher footprint (13 per cent) but it has a proportionately higher energy content to compensate. Diesel engines are typically about 30 per cent more efficient at turning fuel energy into vehicle movement. But if only it were that simple. The downside is that, although they have got much cleaner in recent years, diesels also belch out many more

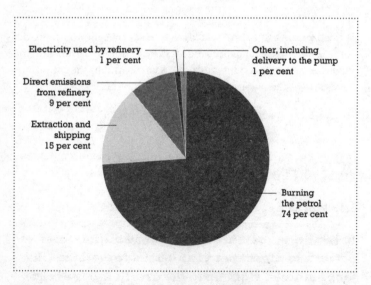

Figure 5.3. The footprint of petrol is more than just the CO_2 that comes off the fuel.

particulates, and this also contributes to climate change (see Black carbon, page 170) as well as causing asthma. Overall, it is hard to say which wins as the environmental vehicle fuel.

Biodiesel deserves a mention as a controversial option, full of problems but also with some potential for the future. The first thing to say is that using land to grow fuel rather than food puts pressure on the world's forests, and chopping trees down already accounts for 18 per cent of global emissions (see Deforestation, page 154). With a fast-growing world population, land is likely to feel increasingly scarce in future. The second negative point to make is that considerable emissions are involved in the growing of the fuel crop and the process of turning it into fuel. Some people even think this can even add up to more than the emissions from fossil fuels. On the plus side for biofuels is the potential to create them from unavoidable agricultural waste and the prospect that future technologies may be able to create them efficiently from algae. Overall, biofuels might one day be a useful part of the solution, but they are not likely to be a magic wand.

1 kg of rice

2.5 kg CO₂e efficiently produced
4 kg CO₂e average
6.1 kg CO₂e inefficient production with excessive use of nitrogen fertiliser

A typical kilo of rice causes more emissions than burning a litre of diesel.

Rice deserves a place in your consciousness not only as a food on your table but as an important piece of the global jigsaw. It provides 20 per cent of the world's food energy in exchange for 3.5 per cent[12] of its carbon footprint (Figure 5.4). Its global footprint is just a fraction less than that of cement production (see page 74). Europeans and Americans get just 1 or 2 per cent of their food energy from this

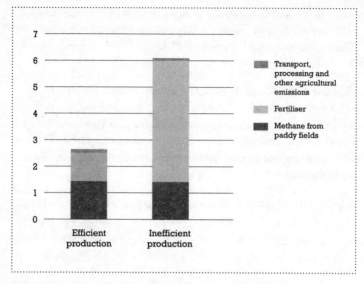

Figure 5.4. The carbon footprint of 1 kg of rice (in kg CO_2e)

crop, but the figure is very much higher in Asia, where 89 per cent of the world's total rice harvest is consumed.[13]

I suspect that plenty of greens will be slightly unsettled to hear that rice, the simplest of foods, is a surprisingly high-carbon staple, much more so than wheat, which is nutritionally similar. That's because of the methane that bubbles out of the flooded paddy fields and the excessive helpings of fertiliser that are all too often applied.

Around the world, 600 million tonnes CO_2e of methane is thought to be emitted from rice paddies, accounting for around 1.2 per cent of the total global footprint and about three times the footprint of all the cement produced in Europe. Even more significant are the 161 million tonnes of fertiliser, mainly nitrogen-based, that are applied to the crop.[14] That's a little over 1 tonne of fertiliser for every 3 tonnes of rice produced. If this is made in an efficient factory and applied sparingly, at well-chosen moments, each tonne applied may only result in 2.7 tonnes CO_2e. If not, the figure could be as high as 12.3 tonnes CO_2e.

I have guesstimated just 100 g CO_2e per kilo of rice for the production of agricultural machinery and the transport of the rice to market. The majority is eaten in its country of origin, and I can't imagine rice ever finding its way into a hothouse or an aeroplane.

If we were to feed the world entirely on food that was as carbon intensive as rice per calorie, what would happen? With smart use of fertiliser, global emissions from agriculture would come in at 11 per cent of the current total, compared with the current 18–20 per cent. But if the worst practice were the norm, agricultural emissions would increase.[15]

It's possible to grow rice without flooding the field and thereby cut out the methane. However, it's harder work (because you have to do more weeding) and you might need more fertiliser, which would mean a trade-off that could end up taking the net carbon impact either way. As with lots of agriculture, we don't fully understand what goes on or what the best options are. It is another important area for research, as the number of human mouths soars ever higher, along with the global temperature.

See also Fertiliser, page 138.

Desalinating a cubic metre of water

Zero CO_2e solar powered (technology permitting)
5 kg CO_2e average, a reverse osmosis plant in Sydney using electricity from coal
23 kg CO_2e inefficient thermal desalination plant

Globally, desalination accounts for perhaps 0.6 per cent of our footprint – and this is rising fast.

Each day the world desalinates around 60 billion litres of water (that's 60 million tonnes or cubic metres), and that figure is currently doubling about every decade.[16] Something like half of the total takes place in and around the dry, oil-rich Middle East but desalination also accounts, for example, for 13 per cent of California's electricity usage and 31 per cent of its gas consumption. Emissions per litre vary hugely depending on the efficiency of the process and the carbon intensity of the electricity used. If a new plant in Sydney is typical of the global efficiency (it uses relatively efficient technology but powers it with electricity from coal), that leaves a global carbon footprint of about 300 million tonnes CO_2e – or something like 0.6 per cent of all global emissions. And that figure is likely to continue increasing rapidly, not least because the world is getting hotter and drier in many regions – a feedback loop of climate change. Spain, which looks set to be one of the countries within Europe most directly affected by changing climate so far,[17] doubled its desalination between 2000 and 2004.

At the high end of the spectrum are inefficient thermal desalination processes. A big improvement on this is reverse osmosis. Various options exist for using spare heat from fossil-fuelled power stations (though we should be careful not to double-count the benefits: in these situations the desalination plant may claim carbon neutrality while the power station claims to be offsetting its emissions by supporting the desalination plant).

At the low-carbon end, Seawater Greenhouse[18] claims to have developed a technique for using solar heat to desalinate water for greenhouse-cultivated crops in arid regions. In theory the desalination itself is just about carbon neutral. I haven't personally investigated the technique but the company has won awards and has some large pilot projects already in operation. I include it here because it is the kind of technology that gives hope in the midst of increasing desertification problems around the world.

Apart from greenhouse gases, another nasty by-product of desalination is the brine concentrate that is returned to the sea, increasing the salinity and messing up marine ecosystems.

A pair of trousers

3 kg CO$_2$e my favourite old nylon travelling trousers
6 kg CO$_2$e my cotton jeans

'Natural' materials may sound greener, but the footprint tells a different story.

My cotton jeans weigh 600 g. Cotton has a footprint of around 7 kg CO$_2$e per kilo.[19] On top of that there is dying, cutting, sewing, an allowance for waste fabric, buckles and zips, transport and so on, which probably takes the total to about 6 kg per pair – equivalent to an 8-mile drive in an average car.

But this figure doesn't tell the whole story. Over the 4 years I suspect I'll get out of them, the footprint of washing and drying them is likely to be several times the footprint of producing them in the first place. My best estimate is four times more (Figure 5.5).[20] It all depends how

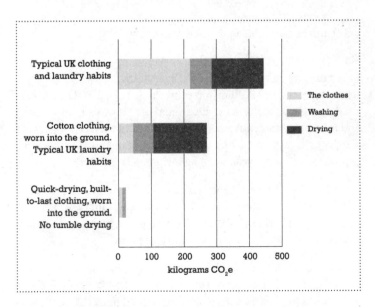

Figure 5.5. The annual carbon footprint of buying and washing clothes.

quickly I get them dirty (quite quickly) and how tolerant I am of the dirt (probably more than most) and how they are washed and dried. But there's no avoiding the fact that the jeans are heavy when wet and they take a lot of drying.

At the other end of the scale is a pair of trousers I've had for 12 years. They have travelled extensively and I've worn them endlessly in the UK, too. They weigh just 250 g and they dry fantastically fast. I can't seem to wear them out. They cost £50, which felt like a lot at the time but now seems a bargain. They are made of some fancy fine-woven nylon. Nylon has a footprint of between 7 and 9 kg CO_2e per kilo depending on the specification[21] – so 12 kg CO_2e per kilo of finished trousers, or 3 kg per pair, is probably about right.

Even if I wear my cotton jeans right into the ground, I can't envisage getting more than 200 days of solid use out of them. That works out at a minimum of 30 g CO_2e per day – or more than 100 g per day once I factor in the laundry. By comparison, my nylon trousers are probably good for 600 days of wear, so that's just 5 g per wear-day, and because they virtually drip dry the laundry aspect probably adds only an extra 6–12 g. All told, then, the nylon trousers are less than one-tenth as carbon intensive as the jeans.

What about the rest of my clothes? If a pair of trousers makes up one-quarter of my daily clothing, and if everything I wore were equivalent to my jeans, my clothing footprint would be 45 kg CO_2e per year for the garments themselves plus around the same again for washing and more than double again if I use a dryer. But if everything I wore were equivalent to my nylon trousers, I could cut my clothing footprint to just 7 kg per year, or 16 kg including laundry.

In reality, the average UK person has a clothing footprint closer to 225 kg per year, or more than 400 kg including laundry, which suggests, not surprisingly, that we are nothing like as good at wearing things into the ground as my scenarios suppose.

All told, if you live in the UK, clothing and textiles will typically make up about 2 per cent of your footprint. And there are broader environmental issues to consider, too. For example, the Aral Sea is drying up partly because of cotton plantations in its catchment,[22] and the clothing and textiles industry produces toxins that find their ways into water supplies.

Here are some tips for keeping the total impact of your clothing to a minimum:

- Buy stuff that is easy to wash and dry.
- Buy stuff that is built to last.
- Wear it and use it until it falls apart, or pass it on.
- Buy second-hand.
- Repair things rather than throwing them out.
- Donate or recycle clothing rather than putting it in the bin.
- Favour synthetic fibres over natural ones.

A steak

2 kg CO_2e a raw 4-ounce beefsteak

A steak has about the same impact as 25 bananas. If you have one per day, that would be more than 700 kg CO_2e per year, equivalent to 1000 average car miles.

Beef is a climate-unfriendly food, coming in at almost 18 kg CO_2e per kilo.[23] About nine-tenths of this footprint comes from the beef farming itself. As we've already seen, using animals to produce food tends to be inefficient compared with eating crops (see page 46), and cows have the added problem that they ruminate, producing enough methane to roughly double the climate-change impact of farming them.

Less widely discussed than the methane are the nitrous oxide emissions, which account for about three-tenths of the footprint of beef farming. This gas is released when nitrogen fertiliser is applied to grass and other fodder crops, and when the grass is silaged. Last, there is the CO_2 itself, at around one-fifth of the farming footprint. This is caused by the tractors, other farming machinery and energy required to make fertiliser.

I'm using the same footprint figure here for all beef, although you could argue the case for attributing more to the most expensive cuts than to the mechanically reclaimed stuff that finds its way into economy burgers. In that sense, offal and processed meat may well be a greener choice than more premium meat products. But however you look at it, food from cows remains towards the top of the carbon spectrum – despite the ongoing debate about whether the footprint of beef and milk can be reduced (see Milk, page 71).

A box of eggs

1.8 kg CO_2e

So a single egg has about the same footprint as four bananas, even before you cook it.

If you tried living entirely off eggs (and survived the cholesterol overdose) you'd meet your calorific needs for around 30 per cent of the total footprint of the 10-tonne lifestyle. This makes eggs less carbon intensive than some animal products but more so than most vegetable-based foods. Figure 5.6 shows that, as with nearly all foods, most of the impact of egg production comes from the farming itself (in this case the rearing of birds and growing of their feed) rather than the packaging or transport.[24] Chickens don't ruminate, so methane isn't much of a problem. But nitrous oxide is the main contributor to the footprint of the final product.

I've based my figures for the farming part of the footprint on a study by Cranfield University. This study suggests that – from the perspective of climate change – organic eggs come out about 25 per cent worse than those from battery farms. This just goes to illustrate that if responding to climate change sends us into a blinkered drive for

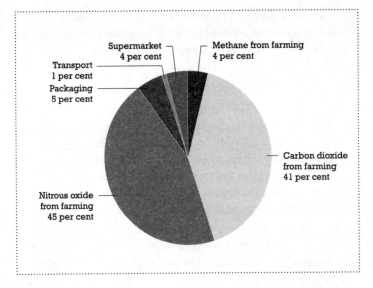

Figure 5.6. How the carbon footprint of eggs (not including cooking) divides up.

efficient production, some other values are going to have to suffer. This book isn't about telling you what values to have, but from time to time it's worth remembering that climate change is not the only issue. If you care about animal welfare as well as climate change, buying fewer eggs but making them organic might be a sensible compromise.

1 kg of tomatoes

0.4 kg CO$_2$e organic loose tomatoes, traditional variety, grown locally in July
9.1 kg CO$_2$e average
50 kg CO$_2$e organic 'on the vine' cherry tomatoes, grown in the UK in March

For the same impact as just 1 kg of the most energy-intensive tomatoes, a 16-stone man could eat his own body weight in oranges.

Shocking! Tomatoes, at their worst, are the highest-carbon food in this book. But at their best, tomatoes are perfectly fine.

At the low end of the scale a high-yield classic variety is grown in the summer with no artificial heat required.[25] The middle and high figures are based on a detailed but controversial study by Cranfield University.[26] The middle figure is averaged across all varieties and times of year. Classic loose tomatoes, the ones that our parents were brought up on, cause only just over half the carbon, at 5.6 kg CO_2e per kilo, whereas 'specialist' varieties (defined here as cherry, plum, cocktail, beef and others) come in at almost 30 kg CO_2e per kilo because the yield is so much lower per hectare, so they need more heat per kilogram.

Perhaps Cranfield's most unsettling finding was that when heat from fossil fuels is required, organic is the highest-carbon option, again because the yield was thought to be lower. At the high-carbon end, therefore, with a staggering 50 kg of greenhouse gas per kilogram, are out-of-season organic, cherry tomatoes sold 'on the vine'.

So, tomato lovers concerned about climate change would do well to stick to the season (July to October in the UK) and to favour classic varieties, sold loose. In the winter it makes carbon sense to stick to tinned tomatoes, but if you do want to buy fresh tomatoes outside the local growing season, it is almost certainly preferable to buy them from Spain or another warmer place rather than choose local versions produced locally in heated greenhouses.

1 kg of trout

5.9 kg CO$_2$e tinned
6.9 kg CO$_2$e fresh from the shop or frozen

I've chosen trout for my example even though I'm not sure I've ever seen it in a tin. That's because up to the point of slaughter it has clocked up a carbon footprint that's fairly average for fish. At this level, fish comes out as a carbon improvement on beef and lamb – and it's healthy, too. But before you rush out and switch your diet over to eating the stuff seven days a week, bear in mind that many of our fish stocks are getting dangerously depleted and if we all switched over from ruminant meat we'd probably wipe out global fish stocks in a decade.

I hesitate to mention that in the studies I looked at, overfished cod comes out with a slightly lower carbon footprint than salmon.[27] Once again, then, we each have to balance up carbon emissions with other concerns. The list in Table 5.2 will help shed some light on the carbon part of the puzzle. For the fish-stocks part, check out the

Fish	Carbon footprint (kg CO$_2$e per kilo)
Mackerel, fresh fillet (at the factory gate)	0.5
Herring, fresh fillet (at the factory gate)	1.3
Unprocessed shellfish (freshly landed)	2.6
Unprocessed general fish (freshly landed)	2.6
Cod, fresh fillet (at the factory gate)	2.7
Cod, frozen fillet (at the factory gate)	2.8
Cod, fresh fillet (at the checkout)	2.8
Cod, fresh fillet (at the factory gate)	3.0
Cod, frozen (at the checkout)	3.2
Trout, frozen fillet (on leaving the slaughterhouse)	4.1
Flatfish, fresh fillet (at the factory gate)	4.7
Frozen filleted fish	6.5
Shrimp, peeled and frozen (at the factory gate)	10.1

Table 5.2. The carbon footprint of different seafoods according to various studies[28]

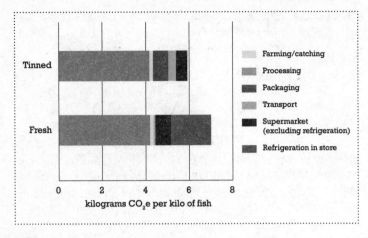

Figure 5.7. The carbon footprint of tinned and fresh fish.

Marine Conservation Society's handy fish-purchasing guide at **www. fishonline.org**.

When tinned fish are compared with fresh stuff over the counter, the refrigeration is a bigger deal than the tin, so fresh fish come out slightly higher in carbon (Figure 5.7). But then the fresh version is 100 per cent fish, with no added oil, and it usually tastes better, too. Fresh fish has a similar footprint to frozen up to the time when you buy it. But the footprint of the frozen version could carry on going up for months in your freezer if you forget it is there.

Leaving the lights on

90 kg CO_2e a low-energy bulb for 1 year
500 kg CO_2e a 100 watt incandescent bulb for 1 year

By 'incandescent bulb' I mean the old fashioned kind with the glowing tungsten wire. In the UK it is now illegal for shops to buy them, so they will soon be museum pieces – at last. Leaving a light on for a whole year might sound extreme, but having an average of one bulb turned on unnecessarily at any one time is almost certainly quite common. (In my office building, which is just 3 years old, and oddly rated environmentally 'Excellent' by the Building Research Establishment, the light is permanently on in the shower. There is simply no off switch.)

As the figures above show, low-energy bulbs have the potential to save an enormous amount of electricity. However, efficiency alone won't bring about a low-carbon world because the less costly something becomes, the more we tend to use it – so the result can end up being more consumption. In the case of lighting, this translates to 'I've left a few lights on but it's OK because they are low-energy ones.' There's also the fact that the money we spend on bills will end up being spent elsewhere – a cheap flight, perhaps (see Rebound effect, discussed within The world's data centres, page 161).

Like any form of electricity wastage, leaving the lights on is one of the cheapest ways of trashing the planet (see page 200), though the precise impact depends on where you live. I've based the figures here on a typical UK energy mix, but if you live in Australia the footprint is about 60 per cent higher. Some people might argue that if you live in France, it's OK to leave the lights on because it mostly comes from low-carbon nuclear power, but in my analysis that doesn't stack up (see Unit of electricity, page 55).

Finally, there's no truth in stories you may have heard that the act of turning a light on uses the same energy as leaving it on for half an hour.

1 kg of steel[29]

0.42 kg CO_2e recycled general steel
2.75 kg CO_2e virgin general steel
6.15 kg CO_2e virgin stainless steel

So virgin general steel has about three times the carbon footprint of cement (or, for that matter, porridge made from half milk and half water) per kilo.

These figures are for steel in its raw form at the foundry gate. In other words they do not take account of any additional emissions that might be required to transport it to wherever it will be used or turning it into something useful like a car or a part of a house. The value of recycling immediately becomes evident, because recycled steel has less than one-sixth of the footprint of its virgin equivalent.

Another key factor is the country of manufacture. This can make a big difference, for three reasons:

- Steel manufacture requires electricity, and the carbon intensity of this varies from country to country depending on how it is generated.

- The amount of energy used depends on the efficiency of the steel plant, with less developed countries often having less efficient manufacturing.

- A final small consideration is that if the steel is made a long way from its final market there will be an additional shipping impact.

A report for the UK government[30] estimated that the emissions associated with the manufacture of 1 tonne of steel in China were typically double those for steel made in the UK, whereas production in the US carried only two-thirds of the footprint and Denmark only half (Table 5.3). India came out worse than China, and Nigeria is worse still – at seven and a half times more carbon-intensive than for UK production.

Country	Carbon emissions
Australia	83
Austria	150
Brazil	117
Canada	133
China	200
Columbia	400
Czech Republic	183
Denmark	50
France	100
Germany	133
India	333
Indonesia	333
Iran	67
Japan	100
Korea (South)	50
New Zealand	300
Nigeria	750
Norway	367
Pakistan	217
Poland	150
Russian Federation	233
Slovakia	167
South Africa	317
Sweden	100
Ukraine	200
UK	100
US	67
Venezuela	283

Table 5.3. Carbon emissions for steel produced around the world (per tonne of steel, as a percentage of UK production)

10 kilos *to* 100 kilos

A pair of shoes

1.5 kg CO$_2$e Crocs
8 kg CO$_2$e synthetic
11.5 kg CO$_2$e average
15 kg CO$_2$e all leather

Imelda Marcos's collection of 2700[1] pairs of shoes would have had a carbon footprint of around 30 tonnes, or 3 years' worth of 10-tonne living – assuming, of course, that they had all been typical shoes.

As the numbers here show, shoes vary enormously in their carbon footprint (no pun intended). Just as important is their longevity.

At the low end of the carbon scale are Crocs, the simple and surprisingly durable shoe consisting of just 250 g of expanded EVA and sold without packaging. For these shoes, the raw material comes in at just over 1 kg. The rest is a guesstimate.

The 8 kg synthetic pair is based on a study of synthetic fell-running shoes, made in China but travelling to market by boat. My average figure, meanwhile, is based on the input-output model (see page 195) and a price of £50 per pair. The model tells us that in the typical shoe about half of the carbon footprint is down to materials, around one-quarter is down to energy used in shoe manufacture, 15 per cent is transport, 5 per cent the shoe box and 5 per cent other bits and bobs.[2]

I have estimated the higher figure for all-leather shoes on the basis of the carbon intensity of cattle farming.

Most of our footwear comes from the Far East, although specialist leather might also have had to travel a long way to get there. Shipping is fairly efficient. The big inefficiency in transport comes if a product is air-freighted for speed. This is most likely in high-end fashion, though unfortunately there's no way to be sure as a consumer what has and hasn't been delivered from the country of origin by plane.

1 kg of cheese

12 kg CO_2e hard cheese

That's about 3 kg CO_2e for a big 250 g block from the shop – equivalent to a 4-mile drive or a massive 12 kg of carrots.

It takes about 10 litres of milk to make 1 kg of hard cheese, adding up to a considerable carbon footprint that's higher than that of many meats. The message is clear, then: going veggie doesn't reduce your impact if you simply swap meat for cheese. Neither will it save you money or make you healthier. Perhaps the best advice if you're keen to reduce the climate impact of your diet is to think of cheese as a meat and therefore a treat. Many people will also improve their life expectancy by cutting back somewhat.

However much cheese you eat, there's an easy carbon win by keeping waste to a minimum. That means buying only what you think you'll actually get through and also avoiding binning hard cheese just because it's showing a tiny sign of mould. This is perfectly safe according the US Food Safety and Inspection Service, which must surely be among the most cautious groups around:

Discard any soft cheese showing mould. For hard cheese, such as

Cheddar, cut off at least 1-inch around and below the mould spot (keep the knife out of the mould itself). After trimming off the mould, the remaining cheese should be safe to eat. Re-cover the cheese in fresh wrap and keep refrigerated.[3]

As for which hard cheese to buy, the most sustainable types probably come from cows that have grazed almost exclusively on rough pasture that couldn't have been used for crops – though of course that information isn't generally available in the shops.

Note that which country or area the cheese has come from doesn't matter much when set against the impact of the milk production (see page 71). Hence the easiest way to reduce the carbon footprint of your cheese is to opt for soft cheeses, because these require less milk to produce.

A congested commute by car

22 kg CO_2e five miles of crawling each way in an average car
Every working day for a year would be 4.8 tonnes CO_2e more than flying from London to Hong Kong and back

A congested drive can cause three times the emissions of the same drive on a clear road.[4]

Driving in queues very roughly doubles your fuel consumption per mile. However, that's only half of the story. By adding your car to the mass of ugly, belching motors, you also make a lot of other people queue just a little bit longer. It turns out, via a bit of simple queuing theory,[5] that the extra emissions you force everyone else to produce (when you add them all together) is about equal to the extra emissions that you produce yourself as a result of having to queue instead of being able to drive straight through. In other words, if your journey is congested, by choosing to do it you cause about three times more emissions than you might expect.

The queuing theory logic also works for the time that gets wasted. If you make the assumption that the journey is many times longer than it would be if there were no traffic, then the time you waste in the queue is about equal to the sum of the extra time you make everyone else waste. In other words, the hassle and anguish that you experience is equal to the hassle and anguish that you inflict. So when deciding whether to drive through a busy area at rush hour, picture your own pain and double it.

All of this adds to the case for travelling by bike, bus, train, foot or lift-share wherever possible. It's also a useful reminder that all motorists should treat cyclists with the respect they deserve for helping to cut everybody else's journey time.

Where you must drive in busy conditions, do your best to minimise stops and starts – both your own and everyone else's. A steady slow stream of traffic is more efficient than a faster but less steady one unless the stops are so long that everyone can turn their engines off. One good tip is to think about what to do when two lanes merge: to reduce emissions, ease your speed down, merge gently and in good time, and allow others to do likewise. In theory at least, two lanes travelling at 50 mph can carry about the same traffic as 3 lanes travelling at 70 mph, assuming that everyone leaves a safe stopping distance between them and the next vehicle. This is because slower cars need less distance between them.[6] Jeremy Clarkson and I don't agree too much of the time, but one point of common ground is that it's good to minimise the use of brakes on the motorway if you can. And when you overtake, put your indicator on in good time too, so no one else has to brake either.

A night in a hotel

3 kg CO_2e low-carbon scenario
25 kg CO_2e £70 spent on dinner, drinks, bed and breakfast in a hotel with average eco-credentials
60 kg CO_2e high-carbon scenario

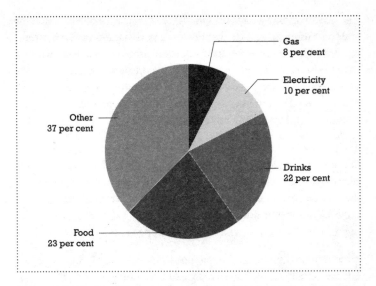

Figure 6.1. The 42 million tonne carbon footprint of the UK's hotels, pubs and catering industry.

For my high-carbon scenario I have chosen one of those hotels where the TV and six lights are already on when you walk into your room. The room itself is too hot and you cool it by opening the window even though the radiator is on. There is a swimming pool, with air-conditioning. You order beef or lamb for dinner and it arrives with baby vegetables air-freighted from Peru. There is too much for you to eat. For pudding you have strawberries even though it is winter. In the kitchens, half of the food cooked is thrown out at the end of the night. You stay one night, finding your way through three towels as well as your sheets. You have a full English breakfast in the morning, giving the paper you ordered a quick glance before leaving it on the table (from where, surely, even in this hotel, it goes for recycling).

The low end of the scale could be a large, very well run hotel or, more likely, a simple bed and breakfast with thoughtful owners. If you stay a few nights your sheets and towel aren't changed unless you ask. The room is comfortable and you can adjust your own heating. You opt for a low-meat-and-dairy meal with seasonal vegetables and you get to choose how much goes onto your plate. Leftovers end up in the next day's soup. You have something like cereal or muesli, fresh fruit and toast for breakfast. There is a selection of papers shared between guests (with the added advantage that you get to browse several if you have time). What you are paying for is a more personalised service in which you can have what you require without it being thrown at you just in case. The difference in carbon footprint between these two scenarios might be as high as a factor of 20.

The British clock up 42 million tonnes of emissions through their use of hotels, pubs, cafes and restaurants (Figure 6.1). That's nearly 5 per cent of the national carbon footprint. What the British drink when they are out has almost as much impact as what they eat, and both of these have a bigger footprint than the energy used by the establishments where the eating and drinking happens.

As a rule of thumb, the hotels, pubs and catering industry in the UK has a footprint of about 400 g CO_2e for every pound you spend. Roughly speaking, this seems to be true whether is it food, drink or accommodation that you are buying. However, this is just a general figure and the footprint certainly doesn't have to go up or down with the price. Indeed, there is a lot that the carbon-conscious consumer can do to keep emissions down, simply by spending money in establishments that think about the issues.

When eating out, look for seasonal fruit and vegetables, and choose places where the lower-meat and lower-dairy options are cooked with at least as much passion as anything else. The restaurant should be taking steps to minimise food waste both on your plate and behind the scenes. In a hotel, look for good energy management, minimisation of laundry and a general sense of care with resources. In a pub, look for local cask beer.

For any hotels, pubs or restaurants seeking to understand their carbon footprint, a colleague and I have built and tested a carbon calculator especially for tourism businesses and have made it freely available online.[7]

A leg of lamb

38 kg CO$_2$e a 2 kg joint at the checkout

For the same carbon footprint, you could have a bowl of porridge (made with half milk, half water) every day for 4 months.

Lamb comes in with a carbon footprint of about 17 kg for each kilo produced at the slaughterhouse. Transport, basic processing, refrigeration and a little bit of packaging each add a little bit, so that by the time the meat reaches the checkout the footprint has increased by about 10 per cent. You will add a similar amount again by the time you have picked it up from the shop, put it in the fridge and cooked it, taking the overall carbon impact to more than 20 kg per kilo.

The issues surrounding sheep are very similar to those relating to cows (see Steak, page 95, and Milk, page 71). Like cows, sheep ruminate, releasing large quantities of methane. And just as with beef farming, the exact impact of different types of sheep farming is complex and only partly understood. Hill farmers can claim that they are putting otherwise unproductive land to use. Some also claim that putting sheep on the hills helps the soil to absorb carbon from the air. Counterarguments are that hill-farmed sheep are inefficient, that they spend too much energy wandering around, eating low-energy food and keeping warm and that therefore they burp more methane per joint of meat than their lowland counterparts.

It seems probable that, from a broad sustainability point of view, hills are the best places to have sheep. But ultimately only one thing is clear: a low-carbon world is going to have to involve less lamb. The typical footprint of this meat is even higher than that of beef. The low-carbon choice is to think of lamb as a treat and to eat less of it.

A carpet

76 kg CO$_2$e thin polyurethane carpet with thin underlay,
4 m × 4 m
290 kg CO$_2$e the same area covered in thick wool,
polypropylene or nylon with generous underlay

If you have 50 m^2 of top-end carpet in your house, that could add up to 900 kg of embodied emissions – equivalent to a burger a day for a year.

Provided you get full wear out of them, some carpets may well pay for themselves in carbon terms, by improving your insulation. However, if you are in the habit of moving house every couple of years and insisting on stripping out everything that was there to replace it with styles more to your own taste, then carpets, along with other soft furnishings, could be a significant chunk of your total carbon footprint.

Table 6.1 gives some figures for the footprints per kilo of fabric materials, based on studies of European production.[8] In the UK most textiles come from developing countries, not least China, where industry is typically a lot more carbon intensive. For Chinese production you can probably mark most of these up by a factor of two

Carpet type	Carbon footprint (kg CO$_2$e per kilo)
General	3.89
Felt underlay	0.96
Nylon	5.43
PET (polyethylene terephthalate)	5.55
Polypropylene	5.03
Polyurethane	3.76
Wool	5.48

Table 6.1. Carbon footprints of carpet types[9]

or three on the basis that the factories tend to be less energy-efficient and the electricity they use is also more carbon intensive per unit because so much of it comes from coal-fired power stations. I'm not writing this out of some kind of protectionist instinct, just presenting the facts as I see them. I'd like to see China develop – but not at any cost.

To give a sense of what the numbers mean in practice, typical weights are 1–1.5 kg per square metre for underlay and 1–3 kg per square metre for the carpet itself. This puts the overall footprint in the region of 5–18 kg per square metre.

Using a mobile phone

47 kg CO_2e a year's typical usage of just under 2 minutes per day
1250 kg CO_2e a year's usage at 1 hour per day
125 million tonnes CO_2e global mobile usage per year

A minute's mobile-to-mobile chatter comes in at 57 g,[10] about the same as an apple, most of a banana or a very large gulp of beer. Three minutes has a similar impact to sending a small letter (written on recycled paper) by second-class post.

Mobile phones cause a fairly tiny slice of global emissions, but if you are a chatterbox using your mobile for an hour each day, the total adds up to more than 1 tonne CO_2e per year – the equivalent of flying from London to New York, one way (economy class).

Indeed, the footprint of your mobile phone use is overwhelmingly determined by the simple question of how often you use it. One estimate for the emissions caused by manufacturing the phone itself is just 16 kg CO_2e,[11] equivalent to nearly 1 kg of beef. If you include the power it consumes over two typical years (that's about how long the average phone remains in use, even though most could prob-

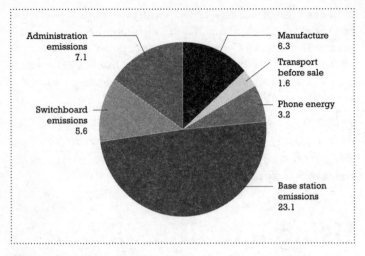

Figure 6.2. The 47 kg annual carbon footprint of mobile usage, based on a Nokia N7600 phone used for 2 minutes per day and replaced after 2 years.[13]

ably last for 10 years) that figure rises to 22 kg.[12] But the footprint of the energy required to transmit your calls across the network is about three times all of this put together, taking us to a best estimate of 94 kg CO_2e over the life of the phone, or 47 kg per year (Figure 6.2).

In 2009 there were 2.7 billion mobiles in use: nearly half the world population has got one. On this basis, mobile calls account for about 125 million tonnes CO_2e, which is just over one-quarter of a per cent of global emissions.[14]

If you want to reduce the footprint of your communication habits, texting is a much lower-carbon option (see page 11). Landlines offer carbon savings, too, because it takes about one-third of the power to transmit a call over a fixed landline network than it does when both callers are on a mobile.[15]

It took a lot of digging to get data for this entry. In the end I was pleasantly surprised that there is some reasonably sensible looking analysis out there. Nevertheless, now feels like a good time for another reminder that all footprint estimates contain considerable uncertainty, and some even more so than others.

Being cremated

80 kg CO$_2$e

That's less than one ten-thousandth of your life's carbon footprint.

My advice would be to treat yourself on this one occasion to whichever method takes your fancy. This is the one time when it is too late to start worrying about your carbon footprint. And anyway you have already done the most carbon-friendly thing possible. However, this book is about doing the numbers, so here goes.

The *Guardian* reported that the average cremation uses 285 kilowatt-hours of gas and 15 kilowatt-hours of electricity, and I'm going with that. I have not included emissions from your own flesh, because your body was only ever a temporary carbon storage device.* On top of the carbon, cremation sends significant amounts of mercury into the atmosphere.

Burial sounds like a more climate-friendly solution, but browsing blogs on this subject (yes, there really are people who write these) I found a clergyman who reckons that burial turns out 10 per cent higher carbon once you take account of cemetery maintenance for the next 50 years. I have managed to resist checking these sums myself.

A sea burial sounds ideal if you don't mind the possibility that some of your loved ones may be heaving over the side when they are supposed to be paying you their last respects. There are usually legal problems with this method, however.

---

* If we start getting into this, losing weight will become a source of guilt. Perhaps even a bit of reverse psychology will kick in and alleviate the western obesity pandemic.

If what you most want to do is send a final eco-message to the world, the best answer I know of is to be dressed in easy-to-rot clothing and put in a wicker coffin. It is possible to be buried in woodland with the idea that your remains will become trees – a lovely idea, though if everyone tried this we might run out of room.

100 kilos *to* 1 tonne

London to Glasgow and back

53 kg CO$_2$e banana-powered bike
66 kg CO$_2$e coach
120 kg CO$_2$e train
330 kg CO$_2$e small efficient car
500 kg CO$_2$e plane
1100 kg CO$_2$e large four-wheel drive

All these scenarios are based on one person travelling the 405 miles each way on their own. I've based the figures for the small efficient car on my own Citroen C1 travelling at a steady 70 miles per hour and getting 55 miles per gallon, which I know is realistic. The four-wheel drive, meanwhile, is based on a Land Rover Discovery doing 19 miles to the gallon. If it goes above 70 miles per hour or puts the air-conditioning on, its impact will be higher still.

For all the road vehicles, the exhaust-pipe emissions make up about half of the footprint. About one-third lies in the manufacture and maintenance of the vehicle itself, and the remaining one-sixth is down to the supply chain of the fuel (see Petrol, page 87). I've assumed that you keep to the speed limit and look after your car with about average care.

The bike is the outright winner if you can afford the time, you are careful about what you eat (see Cycling, page 23) and you don't have a headwind. Of the more practical options, the coach comes top, with a footprint more than 15 times smaller than the gas guzzler. One reason that the coach beats the train is that they travel more slowly, which is significant because the energy needed to overcome air resistance goes with the square of the speed. Another reason is that although a coach is heavy, the weight per passenger is much less than it is for a train (see page 40).

Some analyses that I've seen put a train ticket and a solo drive closer together in carbon terms. But I'm suspicious of these claims because the embodied emissions of the car per passenger mile are often ignored or underestimated. Whatever the precise difference (and it will of course vary widely depending on the particular vehicles), the train also lets you get some work done, read a book or sleep instead of arriving at the other end stressed and frazzled.

The plane could actually be better than driving if you have the wrong kind of car. (My sums are based on flying economy class.) But please don't take this as an advert for flying: it's just a reminder of quite how carbon-profligate some road vehicles are.

As soon as there are more people on the trip, of course, cars become a lot more efficient. If we load the whole family into my C1, along with everything for a week's holiday and put bikes on the back (it is possible, but only just), the fuel consumption goes down by at least 10 per cent. But the emissions per passenger fall so low that we'd be better going that way – in carbon terms, at least – than all travelling by train.

When it comes to both speed and safety, trains and planes win. When you are calculating how much of your life will be taken up by the journey, my back-of-the-envelope calculations tell me that a driver with a fairly typical life expectancy should add about 2 hours each way to the car journey time to take account of the 1 in 200,000 chance that they will lose the rest of their life in a crash.[1] If you are in your twenties and in good health you might want to call it 3 hours. This is a very significant chunk to add on to the expected journey time of 7 hours.[2] For trains and planes the average loss of life expectancy through injury or death is vanishingly small, despite the lavish media coverage that any crash does get. I'm sad to have to report, for the sake of even-handedness, that the bike will lose hands down on safety grounds unless you are careful with your route choice.

A common myth is that huge four-wheel-drive guzzlers are safer for their occupants. This is generally not true. They are, however, more dangerous for everyone else on the road.

Christmas excess

4 kg CO$_2$e per adult low-carbon scenario
280 kg CO$_2$e per adult UK average
1,500 kg CO$_2$e per adult high-carbon scenario

A full-on Christmas could cost you a couple of months' worth of 10-tonne living.

I said at the beginning that this book was about picking your battles. Christmas has got to be a good place to go looking, even if it might entail breaking a few habits and engaging in some delicate family negotiations. For most of us there is a golden opportunity here to escape some mindless consumerism, stress and perhaps even debt.

In my numbers I have only included unwanted presents, wasted food, avoidable travel, fairy lights and cards. Clearly it's not a complete list, but enough to give a flavour. The numbers are per adult and are based on three scenarios, none of which is intended to be ridiculous.

The average adult spends a massive £440 on presents, of which 20 per cent will be totally unwanted.[3] There will also be a lot of 'partly wanted' middle ground, so I've assumed an average 'wantedness factor' of 50 per cent for all presents. In the festive season we spend about £150 more than usual on food, and I've allowed one-third for waste, thinking that this will be slightly higher than it is in the rest of

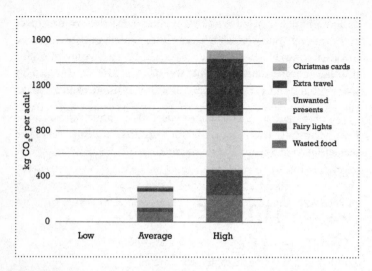

Figure 7.1. The footprint of Christmas waste in the three scenarios.

the year because of the 'Oh-no-not-turkey-again' effect and the fact that the big meals tend to keep coming over the whole period long after most of us have reached our 'wafer-thin mint' threshold.[4] The fairy lights burn through about 45 kilowatt-hours. The average adult posts about 20 cards, with most of the footprint coming from the delivery not the paper. We typically travel 50 miles each above what we would do anyway, and it is generally by car.

In the high-carbon scenario, you spend £1000 on presents (yes, that feels extreme to me, too, but it's only a little over double the average). Sadly, in this scenario the 'wantedness factor' turns out to be just 30 per cent because you are even worse than me at choosing presents. People are too embarrassed to tell you or to sell them, so they gather dust or even get sneaked into landfill. You decorate your house with a wild lighting display that doesn't use LED bulbs. You post 200 rather large cards. You also clock up 500 miles on a tour of relatives in a thirsty car.

I think the low-carbon scenario could be at least as festive and a lot less hassle. The food is great but none gets wasted. You might eat a bit too much, but you make up for that over the coming months, so it's not additional. Your presents are thoughtful but not necessarily expensive. You encourage people to be honest in their reaction and you've kept all the receipts. You have LED fairy lights. You stay at home and you send cards only to a few people that you haven't seen for ages and with whom you really don't want to lose touch. You video-Skype your distant relatives and make plans to see them properly another time.

Some friends of ours spread the word that only children were going to get presents worth more than a strict limit of £1. They asked everyone to reciprocate, packing any cash saved off to the charity of their choice. Both giving and receiving became an exchange of gestures and altogether more fun.

Insulating a loft

350 kg CO_2e outlay for a three-bedroom house
35 tonnes CO_2e payback over 40 years

The payback of insulating a loft can be a remarkable three and a half years' worth of 10-tonne living.

My calculations are based mainly on figures produced by the Energy Saving Trust[5] and assume you are adding 270 mm of rockwool insulation to the previously uninsulated loft of a three-bedroom house. According to the EST's figures you save 800 kg CO_2e per year, but I've rounded this up to 880 kg to take account of fuel supply chains that I know they don't include.

The embodied energy of the insulation material pays back in less than six months and is good for at least 40 years. You will therefore save about 35 tonnes of greenhouse gas.

In terms of money, even without a government grant, you'll get payback on your £500 investment in 4 years, even when a 10 per cent discount rate is applied. In other words, the decision to insulate your loft tomorrow will save you £900 on top of paying back your outlay compared with investing the money in a bank account with a 10 per cent interest rate. (See page 134 for more on discount rates.) In other words, it's a no-brainer. In the UK, the EST may well offer you a 50 per cent grant, too, which makes it a no-brainer even if you are suspicious that they may have been optimistic with their numbers.

Table 7.1 gives a detailed breakdown for the scenario discussed so far and also for someone increasing their insulation from 50 mm to 270 mm. This is a good move, too, but only if you care about the carbon savings or can get a grant. If you are just in it for the money, and you apply a discount rate, then I don't think you ever quite get it back again. However, at just £5 per tonne, the CO_2e saved improving your existing insulation is still a hugely cost-effective way of investing in a lower-carbon world.

The EST's calculations that I've used here are based on the assump-

	From no insulation to 270 mm	From 50 mm insulation to 270 mm
Cost without a grant	£500	£500
Annual payback	£150	£150
Embodied emissions in the material[6]	380 kg	380 kg
Annual carbon saving (including fuel supply chains)	880 kg	880 kg
Financial payback period (with 10 per cent discount rate applied)	4 years	Never quite makes it
Payback over 40 years (with 10 per cent discount rate applied)	£900	−£50
40-year carbon saving	35 tonnes	10 tonnes
Profit or cost per tonne of carbon saved	Net profit of £70!	Net cost of £5

Table 7.1. Insulating the loft in a three-bedroom house without a government grant: the money and the carbon

tion that rather than cashing in on all the financial and carbon savings that would be possible if you kept your home at the same temperature that it used to be, you will in fact allow your home to be warmer once it is insulated. In other words the sums here assume that you will lose some of the available savings in exchange for a warmer and perhaps more comfortable home.

Various types of loft insulation are available: you can get the standard synthetic kinds as well as varieties from sheep's wool, paper and a range of other options. Some of these sound good, but you should only choose them if you are 100 per cent convinced that there is no compromise on performance or the longevity. Those are the priorities.

A necklace

Zero CO_2e handed down or made from driftwood and seashells
200 kg CO_2e £500 worth of new Welsh gold
400 kg CO_2e £500 worth of gold and diamonds sweated out of mines in developing countries

Who would have thought that something so small could have such an impact! But think about it for a moment and it makes sense: gold and diamonds are precious precisely because it takes effort and sweat to extract them.

At the bottom of my scale are items for which the value is in the art and not the materials. Also at the low end of our scale is a piece of jewellery that has been passed on or reforged from an existing item. The carbon impact here is simply from the energy required to melt it down.[7]

To arrive at my ballpark figure for the carbon footprint of jewellery – 400 g CO_2e per pound spent – I have once again used the technique of working out the carbon footprint of an industry and dividing it by that industry's total output. The same model that we used to get the overall figure can give us an idea of where that footprint comes from.

Not surprisingly, it turns out to be attributable to the extraction of metals and minerals, such as gold and diamonds.

For my 'average' example, I have chosen a necklace from virgin Welsh gold, simply because although it has been mined especially for you, this has been done using relatively efficient mining technologies and in a country where machinery tends to be more fuel-efficient than in developing countries. The price of the Welsh gold will reflect the relatively high fuel taxes in the UK, and this reduces the footprint per pound spent somewhat.

At the top end of the scale is jewellery obtained using inefficient technology and cranky machinery. My figure of 800 kg CO_2e per £1000 is simply a crude estimate based on twice the carbon intensity of typical UK industry.[8]

While on the subject of gold and diamonds, chunks of the Amazon are being deforested in the pursuit of gold. Poor people in developing countries are being exploited in the production of both gold and diamonds. Is it worth it? Can it really be a romantic gesture to give someone something that has an embodied footprint of exploitation? Or can there be beauty and elegance without the extravagance?

A computer (and using it)

The machine itself
200 kg CO_2e a simple low-cost laptop
720 kg CO_2e a 2010 21.5-inch iMac
800 kg CO_2e an all-the-frills desktop

Electricity consumption
12 g CO_2e per hour an energy-efficient laptop[9]
63 g CO_2e per hour a 2010 21.5-inch iMac
150 g CO_2e per hour an old desktop machine

Your use of servers and networks

Typically 50 g CO₂e per hour this is the fastest-rising part
of the footprint of computing (see Data centres, page 161)
Add a bit more for any peripherals and the demands you
place on other machines via your use of the Internet

The machine itself

**Even before you turn it on, a new iMac has the
same footprint as flying from Glasgow to Madrid
and back.**

Apple has carried out a detailed life cycle carbon assessment of their
business and their products.[10] This analysis suggests that the company's mid-sized desktop machine – the 21.5-inch iMac – comes in
at around 570 kg CO_2e.[11] However, the devil is in the detail, and the
life cycle approach that Apple has used has a nasty habit of 'leaking'
– missing little bits of the footprint out. The footprint of a computer
comes from the complex mass of activity that has had to go on
throughout the economy in order for minerals in the ground to turn
into machines in the shops. Each component is in turn made of materials and other components, behind each of which lies a whole life
cycle of its own. To trace this by mapping out the different processes
one by one is impossible because the ripple effect is mathematically
endless. You have to miss some processes out, cutting the pathways
short, and the result is a shortfall in your footprint calculations that
is known as 'truncation error'. And it's the reason that I think Apple's
figure is almost certain to be a little on the light side.

The 'input–output' approach of tracing carbon impacts through the
economy by modelling the way in which industries buy and sell from
each other has, for all its generalisations, the huge advantage of not
systematically missing bits out (see page 200). Based on a 21.5-inch
iMac costing £1200 in the shop, input–output modelling gives me a
footprint estimate of about 720 kg CO_2e. Just as expected, that is a bit
higher than the figure produced by Apple's process-based approach,
so it is the one I have gone with.[12]

Apple, on its website, talks about reducing its impact by making machines lighter, but the bulk raw materials are just a small part of the issue. If a laptop were just a lump of plastic, steel and semiconductor, you could get its footprint to below 50 kg CO_2e. The problem is that microprocessors come in at around 5 kg CO_2e for a 2 g chip because of the high-tech process that is involved and the incredibly clean environment that is needed.[13] Apple also talks about reducing packaging; this, too, is good practice but makes a marginal difference in the scheme of things.

It's hard to give guidance on what makes a low-carbon computer because the processes involved in making one are so complex. The guidance we would get from input–output analysis is that the cheaper your machine, the less its footprint is likely to be. This is probably a reasonable rule of thumb, although it may mask the impact of some cheap, carbon-intensive production in developing countries. Another guiding principle could be to choose the produce of a country that has efficient industry and not too much reliance on coal for electricity – but this is tricky because assembly might take place in a country other than the one in which the most energy-intensive 2 gram components are made.

Using the computer

The electricity emissions typically equal the footprint of manufacture after 16,000 hours – that's 10 hours every day for 5 years.

Apple report that the iMac we're talking about consumes 91 watts of electricity in use.[14] They also report that the power supply is 87 per cent efficient so, if I understand them correctly, that makes a total of 105 watts leaving your mains plug. If that is right, in the UK the emissions from use would equal my estimate of embodied emissions after 11,500 hours (that's 7 years at 8 hours per day for 200 days of the year), and by this time the cost of the electricity will probably also have worked out about the same as you paid for the machine. In Australia it would take just 7000 hours because more of the electricity comes from coal (see page 197). In the US it would take 11,000 hours and in China about 8,000. Most people would change their machine before clocking up those hours, so the embodied emissions in the machine are the biggest deal.

However, the sums don't necessarily always work out like that. The iMac is a high-value computer and I have associated that with a relatively high footprint. In addition, some machines are a lot more power hungry. Apple say they have worked on becoming more efficient. Traditionally, laptops consume less than desktops, because it has always been important to conserve battery life. Some, but not all, desktops are catching up. I recently encountered an office full of fairly new HP and Compaq PCs that were burning through 24 watts even when they were switched off. Since they were only on during office hours they were consuming more when turned off than on. The answer was simply to unplug them at the end of the day.

I haven't taken into account the use of peripherals or the activity you might stimulate in other machines around the web through your emails and web searches (see Data centres, page 161). Despite all this, computing can be a fairly low-carbon way of spending time.

To summarise, computing could be a few percent of your carbon footprint. The embodied footprint of a computer is significant and could easily be the dominant factor, so it probably doesn't make sense to buy a new, more efficient machine on carbon grounds; better to make an old one last. However, when choosing a computer, do think about its power consumption, especially if you will use it a lot. Laptops are still usually better in this respect than desktops, but whatever you use, switch it off when not in use and unplug it if that's what it takes to zero the power.

A mortgage

800 kg CO$_2$e per year for £100,000 on 5 per cent interest

That's a whole month of 10-tonne living.

How can a mortgage have a carbon footprint? Surely it just boils down to a few bits of paper or electronic transactions? Look more closely. The bank or building society runs offices, buys computers, sends mail (probably mainly junk; see page 44) and stores data. Its people travel. It outsources everything from cleaners to building maintenance, from design work to corporate lunches, and maybe even still buys in the odd paper clip.

What I am saying is that when you take out the loan you feed the financial services industry along with all its direct and indirect environmental impacts. This is another example of a set of ripple effects across the economy that we can't see and don't stand a chance of counting up one by one. Happily, our input–output model (see page 195) comes to the rescue and produces a ballpark figure of 160 g CO_2e for every pound spent on UK financial services.

If you have a £100,000 mortgage on a 5 per cent interest rate you pay £5,000 per year (plus any actual repayments) and this incurs an annual footprint of the order of 800 kg CO_2e. The same story applies to all loans, and the principle goes wider still. All the intangible services have fairly similar carbon intensities: solicitors, lawyers, accountants, therapists, architects, and so on.

There are two basic lines of attack if you want to cut the carbon. The first is to take out a smaller mortgage and spend the money you saved on something that decreases carbon emissions, such as an investment in an offshore wind farm, a 'save the rainforests' project, or, if you want your neighbours to know what a good person you are, a solar roof. You could stick the money in the bank where it may seem harmless, but even then you may be enabling the bank to lend more to profligate consumers. The other line of attack is to be discerning about the way the mortgage company goes about its affairs. I have based the footprint estimate on general figures for the industry, but actually there will be good and bad practices within it. To begin with, one-tenth of the sector's footprint comes from printing and postage, so supporting a bank or building society that doesn't do junk mail is a good first step. About 30 per cent of the industry's footprint comes from air transport, but I'd be surprised if, for example, the Ecology Building Society, based in Yorkshire, goes in for much of this. They run a simple, lean operation out of eco-friendly premises and make a real effort to walk the walk. If I had to guess I'd put their carbon intensity at less than half of the industry average. Furthermore, their footprint is in the cause of encouraging a sustainable building stock, because they vet their loans by the sustainability of the project and also support lenders in improving their buildings.

The job of choosing between more mainstream lenders is trickier. The most important question is probably to ask yourself how much you *trust* them to take the carbon issue seriously. If the answer is that you don't, then they probably haven't done much to be any better than the industry norm, no matter how much they are talking about it. That is my experience in most of the industries I work in. With my own money, I took the view that the Co-operative Bank was better than most, so I bank there.

1 tonne *to* 10 tonnes

A heart bypass operation

1.1 tonnes CO_2e

That's nearly six weeks of a 10-tonne lifestyle – equivalent to a couple of return flights from London to Madrid.

The carbon cost of healthcare in the UK is around 170 g per pound spent. That makes it a fairly low-carbon way of spending money. And in terms of the quality of life improvements we stand to gain from it, healthcare when we need it must be one of best ways of spending our carbon budget.

That said, a big operation clocks up a big footprint. The typical cost of a heart bypass to the National Health Service is £6324.[1] If we assume that this operation is averagely carbon intensive, that adds up to more than 1 tonne CO_2e.

Overall, UK healthcare has a footprint of 27 million tonnes CO_2e, or just over 3 per cent of the national consumption footprint. Electricity and fuel used by health services accounts for less than one-third. Drugs account for nearly one-fifth. My catch-all 'other' category is nearly one-third of the total, reflecting the variety of equipment and other stuff that is required to keep us healthy. Paper and card

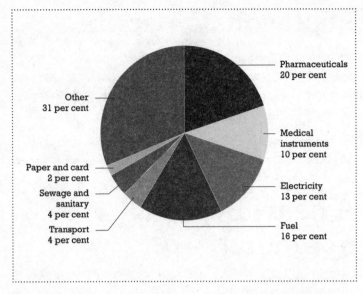

Figure 8.1. The footprint of healthcare in the UK: 27 million tonnes.

surprised me at a massive 2 per cent of the footprint of all healthcare. I'd like to think this is not the stuff that clogs up the filing cabinets of one of the world's biggest bureaucracies but rather the consumables used to keep things clean. So what can we do to reduce the emissions of our healthcare? The best option is to stay healthy, of course. This might involve cycling (safely) or walking more, and thinking about the amount of meat and dairy produce in your diet – all things that will reduce your direct footprint, too, and which are discussed elsewhere in this book. When you do actually need healthcare, be as careful with NHS resources as you would be if you were paying for it directly yourself. But relax in the knowledge that at around 170 g CO_2e per pound it is one of the lower carbon ways for you or your government to spend money.

Photovoltaic panels

3.5 tonnes CO$_2$e producing a solar roof capable of generating 1800 units (kilowatt-hours) of electricity per year
50 tonnes CO$_2$e lifetime saving; that's 5 years' worth of 10-tonne living

Warning: This section contains myth-busting payback calculations that will interest some more than others.

I'm going to do the financial sums and the carbon sums and then put these both together to see how electricity-producing photovoltaic solar panels rate as a cost-effective way of saving carbon.

First, the financial bit. Many governments offer a 'feed-in tariff' to reward individuals who install solar panels on their roofs. In the UK, householders are offered a massive 36.5p per unit generated.[2] This handout is guaranteed for the next 20 years. On top of the feed-in tariff you can still use what you generate yourself (thus cutting the amount you have to buy) or sell it back to the grid to get even more revenue. It's an incredibly generous government handout (especially given the UK's financial situation) and if currently available micro-photovoltaic panels are a viable source of electricity, surely we should all be diving in?

Analyst and author Chris Goodall[3] has done sums on the financial payback from micro-renewables. He estimates that it will cost you £10,000 to get a set of panels installed that is capable of providing you with £1800 kilowatt-hours per year. Once you have taken account of income from the tariff, your sales to the grid and reductions in your grid electricity bill as well as annual maintenance costs, Chris thinks you can make a return of £730 per year. This figure suggests a financial breakeven after 14 years. That sounds fine, but what this is really saying is that provided everything goes to plan you will be exactly as well off as you would have been if you had kept the £10,000 *in a box under your mattress*. Such a simple 'payback period' calculation would be fatally flawed because it would ignore both the fact that you could have done something else with the money, where at the very least you would have got a bit of interest to offset inflation, and the fact that even the surest-looking projects, backed up by manufacturers' guarantees, carry a degree of risk.

More realistic payback sums need to have a way of taking into account the fact that money in your hand right now is worth more to you than the promise of the same amount of money to be paid to you in the future provided that things go well. This can be done by applying a so-called discount rate to the future payback. Applying a 10 per cent discount rate (a fairly sensible figure) is equivalent to saying that you'd be just as happy to have £900 in your pocket now as you would be to have £1000 promised to you in a year's time on the condition that your photovoltaic panel project is still going to plan. Following the same logic, a promise of £1000 in two years' time is worth just £810 to you today and the financial return that you hope to get in your 14th year is worth less than one-fifth of the same money in your hand right now. So, what happens to your solar payback period once a 10 per cent discount rate has been applied? It turns out that you would never get more than two-thirds of your money back, even if your panels lasted forever. (Which they won't. After 20 years they can be expected to be functioning at less than 80 per cent efficiency and after 40 years they will probably have had it.) In other words, don't buy a solar roof purely as a profit-making venture, even with the government's wildly generous feed-in tariff.

But what about the carbon sums? I'll guesstimate that the £10,000 you spend is half on the kit and half on the installation. To give the carbon sums their very best possible chance I'll generously overlook the footprint of installation and use the lowest plausible figure I can take from my input–output model for the manufacture of the panels: 0.7 kg CO_2e per pound spent. That gives the panels a footprint of 3.5 tonnes. If we assume that the electricity generated all replaces output from coal-fired power stations rather than the UK grid average, then the carbon saving per year is about 1.8 tonnes and you'd pay back the carbon in about 2 years. So where does that leave us? After 40 years your net cost (your initial investment minus the paybacks each year with discount rate applied) is still over £3000. The government will have invested £13,000 over the 20 years of the feed-in tariff and (I'm assuming) nothing from then on. Something like 50 tonnes CO_2e will have been saved.[4] That's a cost of £330 per tonne, even worse than a micro wind turbine and dramatically worse than offshore wind.

Are there any reasons to get a solar photovoltaic roof? Perhaps. You might want to invest in a developing technology. Or you might simply want one for fun. If you need to buy things to prove your status in society, solar panels are one of your most carbon-friendly options. We spend billions on mindless junk and flights around the world for that very reason: status. With the panels you can show everyone that you have spare cash but that you also think about the world. Photovoltaic panels can replace the SUV, and you might still be in the vanguard of this trend if you are very quick.

Flying from London to Hong Kong return

3.4 tonnes CO_2e economy class
4.6 tonnes CO_2e average
13.5 tonnes CO_2e first class[5]

Three economy trips are a whole year's worth of 10-tonne living. One trip is equivalent to 340,000 disposable carrier bags.

In other words, for your carrier bags to have the same footprint as just one trip to Hong Kong you would have to go to the supermarket every single day for 10 years and return each time with 93 disposable bags.

A Boeing 747 carrying 416 passengers burns through 116 tonnes of fuel on the 9700 km flight each way. Almost one-third of the total weight on take-off is fuel. As the fuel burns it creates three times its weight in CO_2. But the impact is worse still because high-altitude emissions are known to have a considerably greater impact than their low-altitude equivalents. The science of this is hideously complex and poorly understood,* but there is still a clear case for applying a multiplier to aviation emissions to take account of their extra impact. I have used a factor of 1.9.[6] Aviation is sometimes said to account for between 1 and 2 per cent of global emissions. These statistics ignore the effect of altitude. The proportion is also higher in the developed world, especially in those bits of it, like the UK, that are surrounded by sea. Here, personal flights account for a huge 8 per cent of the carbon footprint of all consumption. That rises to nearly 12 per cent once business flights and air-freight are added on.

In terms of your own lifestyle it might be much less than this. Many British people never fly at all. On the other hand, for some people, flying accounts for the overwhelming majority of their total footprint, and trying to cut carbon in other areas might simply be a misdirection of attention, distracting them from what matters. First-class and business-class tickets are particularly high in impact simply because your seat uses up more of the plane and because by paying more money you provide a greater proportion of the commercial incentive for the flight. It's hard to imagine a low-carbon flying technology coming to the rescue. The physics of flight simply does not allow us to reduce the energy it takes to keep us in the air by more than a few per cent,[7] and for the foreseeable future that energy has to come from fossil fuels. Nevertheless, there are still some efficiencies to be had. One of these is the automation of air-traffic control to replace the current archaic manual system. Humans are woefully unable to calculate optimum flight paths in real time with hundreds of planes in the air at once, all competing for space and time slots. One estimate is that upwards of 9 per cent efficiency improvements are possible.[8]

Ultimately, then, it's hard to avoid the conclusion that most of us

---

* Here is a glimpse of the main issues: The amount of nitrous oxide that a jet engine produces varies with altitude. Its effect on ozone levels also depends on attitude. *And* furthermore the effect of that ozone on climate is altitude dependent. Planes also cause contrails under certain atmospheric conditions, and these are known to make a short-lived but large contribution to the greenhouse effect. The contrails themselves depend on temperature, weather conditions, time of day and, you've guessed it, altitude.

need to fly less. But that needn't make our lives any worse. Make your flights count: go for longer but less often, and do things you really couldn't do at home. For the rest, try local trips, which involve less travel time and therefore more holiday. After all, the experience of getting to an airport, hanging around in a departure lounge and then sitting cooped up for hours are intrinsically rubbish ways of spending time. Also think about *where* you fly to: the closer the destination, the fewer the emissions. One myth is that long-haul flights are automatically more efficient per mile than short-haul because they involve proportionally less time taxiing, queuing, taking off and landing. This isn't necessarily true, because the long-haul flight has to lift more fuel. The truth is that the most carbon-efficient way of getting across the world is in several hops – but not too many.[9] But none of this changes the fact that the further you fly, the larger the footprint.

Of course, the flying conundrum affects companies as well as individuals. I work with a few businesses for whom flying is a key issue. They know it's high in carbon, costly and time consuming. They also know they have always had strong business reasons for doing it. New thinking is required to break out of old habits. Video conferencing may never fully replace human contact, but on the other hand it is a lot cheaper and easier once you are fully conversant with the technology. What is worth more, one face-to-face visit or ten video link-ups?

It is difficult to see a place in the low-carbon world for much air-freighted food (see Asparagus, page 83), let alone durable goods such as clothing. Some garments are air-freighted simply to reduce lead times and cut the cost of stock that is tied up in transit at sea. Air-freight labels are one piece of consumer information that would surely be simple and helpful. Currently these are found on some supermarket fresh produce but nowhere else.

I'm sometimes asked about air freight from developing countries: 'Surely it's good to keep supporting that country by carrying on the trade!' In broad terms, I don't think so. The argument is a bit like saying you should keep the arms trade booming so that people can keep their jobs. Economies need to be powered by people doing things that are useful. Anything else is an unsustainable nonsense.

1 tonne of fertiliser

2.7 tonnes CO$_2$e nitrogen fertiliser efficiently made and
sparingly spread
12.3 tonnes CO$_2$e the same stuff made inefficiently and
used in excess

**A real carbon opportunity: up to half a per cent
reduction in global emissions – it's dead easy and
has no bad side effects.**

Nitrogen fertiliser is a significant contributor to the world's carbon
footprint. Its production is energy intensive because the chemical
process involved requires both heat and pressure. Depending on the
efficiency of the factory, making 1 tonne of fertiliser creates between
1 and 4 tonnes CO$_2$e. When the fertiliser is actually applied, between
1 and 5 per cent of the nitrogen it contains is released as nitrous
oxide, which is around 300 times more potent than CO$_2$. This adds
between 1.7 and 8.3 tonnes CO$_2$e to the total footprint,[10] depending
on a variety of factors.[11] Here's how the science of it goes. All plants
contain nitrogen, so if you're growing a crop it has to be replaced into
the soil somehow or it will eventually run out. Nitrogen fertiliser is
one way of doing this. Manure is another. Up to a point there can
be big benefits. For some crops in some situations, the amount of
produce can even be proportional to the amount of nitrogen that
is used. However, there is a cut-off point after which applying more
does nothing at all to the yield, or even decreases it. Timing matters,
too. It is inefficient to apply fertiliser before a seed has had a chance
to develop into a rapidly growing plant. Currently these messages
are frequently not understood by small farmers in rural China, especially, where fertiliser is as cheap as chips and the farmers believe that
the more they put on the bigger and better the crop will be. Many
have a visceral understanding of the needs for high yields, having
experienced hunger in their own lifetime, so it is easy to understand
the instinct to spread a bit more fertiliser. After all, China has 22 per
cent of the world's population to feed from 9 per cent of the world's
arable land. There are other countries in which the same issues apply,

although typically the developed world is more careful. Meanwhile in parts of Africa there is a scarcity of nitrogen in the soil and there would be real benefits in applying a bit more fertiliser to increase the yield and get people properly fed. One-third of all nitrogen fertiliser is applied to fields in China – about 26 million tonnes per year. The Chinese government believes there is scope for a 30–60 per cent reduction without any decrease in yields. In other words, emissions savings in the order of 100 million tonnes are possible just by cutting out stuff that does nothing whatsoever to help the yield. There are other benefits, too. It's much better for the environment generally and it's cheaper and easier for the farmers. It boils down to an education exercise … and perhaps dealing with the interests of a fertiliser industry.

A person

0.1 tonne CO_2e per year average Malawian
3.3 tonnes CO_2e per year average Chinese person
7 tonnes CO_2e per year world average
15 tonnes CO_2e per year average UK inhabitant
28 tonnes CO_2e per year average North American
30 tonnes CO_2e per year average Australian

Figure 8.2 shows two ways of looking at the emissions per person for various countries: the official 'direct' footprint and my estimate for the 'consumption' footprint. The direct footprints include all the greenhouse gases released inside a country's borders; the consumption footprint is adjusted to take account of imports and exports, giving a total that represents all the goods and services ultimately consumed by a typical person in each country.

For the UK, our direct average footprint of 11 tonnes per head goes up to about 15 tonnes once you include imports and international travel and shipping. I have estimated that for other developed countries the same mark-up of about 4 tonnes per head seems reasonable. In China it works in reverse. About one-third of their emissions go

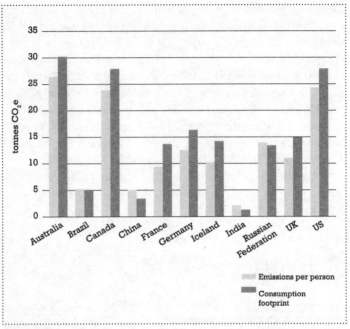

Figure 8.2. Emissions per person and an estimate of footprint per person.

into exported goods, so the footprint of Chinese consumption is only about two-thirds of the emissions that physically come out of the country itself. I've estimated that a similar story applies to India but to a slightly smaller extent.

My estimates of the difference between a nation's emissions and its consumption footprint are very conservative, for one very important reason. The figures are based on the flawed assumption that the overseas production is exactly as carbon intensive as the UK equivalent. In other words, it assumes that if you have your washing machine, your computer and your pair of jeans made in China they will have the same embodied footprint as if they were manufactured in the UK. We know that this is not true and there is a strong argument for using much higher figures for most imports, based on inefficient production and more polluting electricity generation in coal-dependent exporting countries such as China.

10 tonnes *to* 100 tonnes

A car crash

Zero a tiny bump that you can live with
7 tonnes CO_2e a write-off on an empty road
50 tonnes CO_2e a double write-off on a busy motorway

My medium-carbon scenario involves you writing off a £10,000 car on an empty road without damaging anyone or anything else. As a rule of thumb, the embedded emissions in a car are around 700 g CO_2e per £1 of value.

At the high end of my scale you write off two cars, each worth £15,000 and cause a 10-mile tailback of crawling traffic across three lanes of a motorway for 2 hours. If the queue has frequent stops and starts, the 6000 or so cars involved will be unable to turn their engines off and will each emit perhaps 5 kg CO_2e more from their exhaust pipes than they would have if the road had been clear. That tots up to a further 30 tonnes of emissions – which is more than the crash itself.

My sums have not taken account of the footprint of a whole string of other consequences of the crash: the extra burden on the emergency and health services, congestion on surrounding roads, wear and tear on cars, to name but a few. If complex surgery were involved for one or more injured drivers, that could boost the total footprint significantly (see page 131).

It is also interesting to look at the human impact of the crash. Let's say for the sake of argument that one 50-year-old person dies but everyone else is more or less fine. We could say, a little simplistically, that the human impact has been the loss of 30 person-years of life (plus a massive impact on the lives of friends, families and colleagues).[1] As for the broader impact, if the 6000 cars each had a typical 1.6 occupants, about 20,000 person-hours will be spent in the living hell of the queue. That's a bit more than most of us spend awake over a 3-year period, but to stick with round numbers let's say that the human impact is 3 person-years of life lost. Finally, there are the victims of climate change. The extent to which people around the world will be affected by the release of 50 tonnes CO_2e is impossible to quantify, but the exercise of trying might still help us gain perspective. In Chapter 1 of this book I did some rather flaky calculations to arrive at a less-than-robust figure of one life lost for every 150 tonnes CO_2e emitted. If this bears any resemblance to the truth, then the footprint of the crash and the queue together will trigger about one-third of a life lost – say, the loss of another 25 person-years.

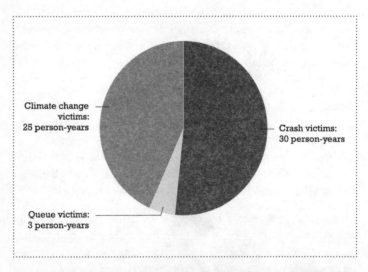

Figure 9.1. The human impact of a car crash in which two £15,000 cars are written off, a 50-year-old dies and there is a 10-mile motorway queue lasting 2 hours. The figures exclude the indirect effects on family and friends.

A new car

6 tonnes CO$_2$e Citroen C1, basic specification
17 tonnes CO$_2$e Ford Mondeo
35 tonnes CO$_2$e Land Rover Discovery, top of the range

So a new gas guzzler could eat up three and a half years' worth of 10-tonne living before you even drive it off the forecourt. (It's not as much as this if you trade in your old car for resale.)

The carbon footprint of a car is immensely complex. Ores have to be dug out of the ground and the metals extracted. These have to be turned into parts. Other components have to be brought together: rubber tyres, plastic dashboards, paint, and so on. All of this involves transporting things around the world. The whole lot then has to be assembled, and every stage in the process requires energy. The companies that make cars have offices and other infrastructure with their own carbon footprints, which we need to somehow allocate proportionately to the cars that are made. When you stop to think about it, the manufacture of a car causes ripples that go right throughout the economy. To give just one simple example among millions, the assembly plant uses phones and they in turn had to be manufactured, along with the phone lines that transmit the calls. It goes on and on for ever.

Attempts to capture all these stages by adding them up individually (the so-called 'process-based' approach to carbon footprinting) are doomed from the outset to result in an underestimate, because the task is just too big. Luckily there's an alternative in the form of the input–output method (see page 195). This approach takes account of all these infinite ripples even if it does rely very heavily on the law of averages. It can give us clues as to the footprint of a car per unit of monetary value and also tell us a bit about how that footprint comes about.

The input–output approach suggests that a motorcar might have a footprint of 720 kg per £1000 that you spend on it. Figure 9.2 shows how this breaks down. The gas and electricity used by the industry

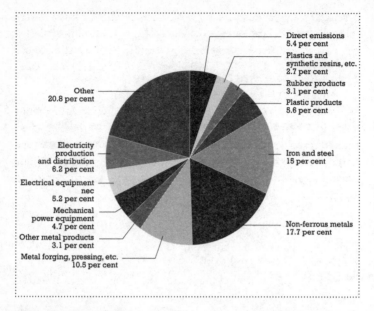

Figure 9.2. The carbon footprint of a car. Gas and electricity used by vehicle manufacturers themselves account for 12% between them. Each of the 'other' slices could be broken down again into yet another pie with a similar story. For example the footprint of the 'plastic products' slice comes partly from gas and electricity used by that industry and partly from everything else across the economy that the plastic products industry spends money on. 'Nec' stands for 'not elsewhere classified'.

itself, including all the component manufacturers as well as the assembly plant, account for less than 12% of the total. The complexity of the pie illustrates just how far and wide the rest of the footprint is dispersed. There is only room to put labels on the biggest slices. This is just the beginning, though. Remember that behind each piece of pie are all the complex supply chains that lie behind that industry in turn. No wonder the process based approach didn't stand a chance.

The upshot is that the embodied emissions of a car typically rival the exhaust pipe emissions over its entire lifetime. Per mile, the emissions from the manufacture of a top-of-the-range Land Rover Discovery that ends up being scrapped after 100,000 miles may be four times as much as comes out the exhaust pipe of my Citroen C1. I have seen plenty of analyses of the question of whether it is a lower-carbon option to keep or to scrap your old car. These almost always rely on process based approaches and therefore underestimate the embodied energy and conclude that you should replace your car far too readily.

My family has a 12-year-old Volvo that does a disgraceful 35 miles to the gallon. If we let it go it would be scrapped, so the embodied emissions can be considered to be written off. Keeping it for the very few journeys that require a big car enables us to do almost all our driving in a little C1. So having a big banger for a second car is a lower-carbon option than having just one new large car.

Generally speaking, then, it makes sense to keep your old car for as long as it is reliable, unless you are doing high mileage or the fuel consumption is ridiculously poor. You can of course boost the life of the car by looking after it. Table 9.1 shows how much lower the total emissions per mile can be if your car lasts twice as long.

When you do eventually replace your car, do so with a light, simple and fuel-efficient model: that way you'll be limiting both the manufacturing and the exhaust pipe emissions. But before you buy, have a look into car-share schemes: you may save lots of money as well as reducing the number of cars that need to be produced. Or if you think you need a car as a personality extension, consider getting a solar roof or a wind turbine instead (see pages 133 and 146).

Model	Price (thousands of pounds)	Embodied emissions (tonnes CO_2e)	Embodied emissions per mile over 100,000 miles (grams CO_2e)	Embodied emissions per mile over 200,000 miles (grams CO_2e)
Citroen C1	£8–10	5–7	50–70	25–35
Prius hatchback	£18–23	13–17	130–170	65–85
Ford Mondeo	£19–28	14–20	140–200	70–100
Land Rover Discovery	£32–48	23–35	230–350	135–175

Table 9.1. The carbon footprint of cars themselves per mile

A wind turbine

30 tonnes CO_2e a 15-kilowatt turbine, installed
500 tonnes CO_2e net savings over a 20-year lifetime

Warning: This section contains payback calculations that will interest some more than others.

Let's have a look at how a turbine might stack up in both cash and carbon terms. A 15-kilowatt turbine is at the biggest end of the micro-renewables spectrum. With a 9-metre diameter and a pole as high as a four-storey house, this is the most efficient form of micro wind turbine, and the sort of thing you could install only if you had plenty of space and money.

According to Chris Goodall's estimates,[2] a 15-kilowatt turbine (that's the maximum output, which is a long way above what the device typically generates), costing £41,000 to purchase and a further £9000 to install, is capable of delivering 25,000 kilowatt-hours of electricity each year if placed on a suitably windy site.

I don't know of any credible studies of emissions of producing and installing turbines, and input–output analysis is likely to be crude even by its own standards in this rapidly changing industry. For this reason, my estimates here are going to be even more broad than usual. However, it is worth having a go. If turbine manufacture is about as carbon intensive per pound sterling of product as other generators and electrical motors, which seems a reasonable assumption, the carbon intensity of manufacture will be around 640 kg CO_2e per £1000 of value. Installation is probably about as carbon intensive as typical construction, at around 380 kg per £1000. That makes the footprint of the installed turbine 30 tonnes CO_2e.

The carbon savings from generation depend on the carbon intensity of the electricity that you're replacing. Let's assume that your generation replaces the coal-fuelled part of the country's energy mix. In other words, if you live in the UK, let's say that rather than replacing typical grid electricity, which comes from a mix of coal, gas, oil and renewables, the effect of your turbine is to reduce the use of coal-fired power stations. That's reasonable because coal is the least preferable source in the electricity mix (see page 55). In this case the carbon saving is roughly 1 kg per kilowatt-hour, so you save 25 tonnes per year and pay back the embodied carbon in just 14 months – a great start.

So how about financial payback? The UK government has recently introduced a 'feed-in tariff' that pays you 24p per unit on top of all the money you save on your own fuel bill and from selling surplus back to the grid at approximately 5p per unit.[3] With all this taken into account you would get back £7250 per year on your investment. That pays back in about 6 years if you forget what else you might have done with the up-front money. However, most economists would apply a 'discount rate' (see page 134) to take account of the fact that it's usually more useful to have money right now than to have to wait for it. If you apply a 10% discount rate (in other words you'd value a firm promise to be given £1000 in a year's time exactly as highly as an offer to receive £900 right now), the payback period becomes 8 years, but it still makes good financial sense. And if you care about the carbon savings for their own sake, it looks like a fantastic move. The carbon investment pays back in just over a year, and every year after that is a 25-tonne carbon saving. (Please don't forget that all these sums rely on your wind turbine having a favourable location.)

So, at face value, the turbine looks like a great idea environmentally and a fairly good long-term investment economically for the person installing it. But there is a crucial perspective missing from the analysis so far. Has the government spent its money wisely? It has invested 24p per unit into your turbine. That works out at a massive £250 per tonne of carbon saved.

My sums tell me that had the government invested its money in large offshore wind farms, instead of subsidising your smaller turbine, they would have broken even after 8 years (without discount rate) or 15 years (with a 10% discount rate).* In other words, the micro turbine works out as a good investment for you, but only because the government spends, and arguably wastes, so much money subsidising it. There are carbon savings but they are far less than could be had if the government put the investment into, for example, offshore turbines.

Suppose that the wind turbine feed-in tariff were to encourage enough micro-turbine installations to reduce direct UK emissions by 1%. That would cost £1.8 billion – in contrast with getting the same effect from macro-renewables, which would ultimately have paid back financially. In this light, although the wind-turbine feed-in tariff doesn't look like the very best way of spending government resources on climate change mitigation, we are talking about investing about 0.075% per year of the nation's GDP to get a 1% reduction in the emissions. In other words, it could be much better but it could be worse. The investment also supports an important developing technology.

There is one extra favourable way of looking at the micro wind turbine, even if it is not the single best way of investing money in cutting carbon. Input–output modelling has told us that it is actually quite difficult to spend money without having a negative carbon impact (see Spending £1, page 59). So if the feed-in tariff encourages people to spend their money on a carbon-reducing technology such as a wind turbine, rather than on carbon-producing goods and services such as a car or a series of overseas holidays, then the reductions in emissions will be greater than my simple sums above have suggested. (See also Photovoltaic panels, page 133.)

---

*A report by Sinclair Knight Merz estimates a £55 billion investment to replace 40% of the UK's electricity with renewable sources, mainly offshore wind, by 2020. That is 130 terawatt-hours per year (130,000 billion units per year). This pays back in 15 years even with a 10% discount rate and valuing generated electricity at just 5p per unit. Sinclair Knight Merz (2008) *Growth Scenarios for UK Renewables Generation and Implications for Future Developments and Operation of Electricity Networks* BERR Publication URN 08/1021, **www.berr.gov.uk/files/file46772.pdf** [accessed October 2009].

A house

80 tonnes CO$_2$e

That's equivalent to five brand-new family cars, eight years of 10-tonne living or 24 economy-class trips to Hong Kong from London.

This figure is for the construction of a brand-new cottage with two bedrooms upstairs and two reception rooms and a kitchen downstairs. Figure 9.3 shows the footprint of the materials. These numbers come from a study I was involved in for Historic Scotland.[4] We looked at the climate change implications of various options for a traditional cottage in Dumfries: leave it as it is, refurbish, or knock it down and build a new one to various different building codes. We looked at the climate change impact over a 100-year period, taking into account

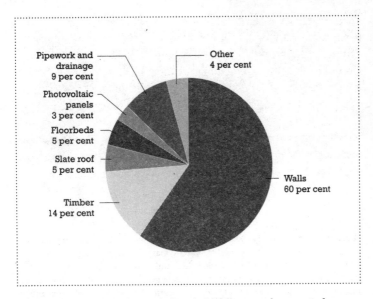

Figure 9.3. The embodied emissions in building a carbon-neutral two-bedroomed cottage. The photovoltaic panels struggle to justify themselves financially.

the embodied emissions in the construction and maintenance as well as living in the building.

The worst option by far was to do nothing and leave the old house leaking energy like a sieve. Knocking down and starting again worked out at about 80 tonnes CO_2e whether you built to 2008 Scottish building regulations or to the much more stringent and more expensive Code for Sustainable Homes Level 5 that demanded 'carbon neutrality'. For all the new-build options, the up-front emissions from construction work are paid back by savings from better energy efficiency of the house in 15–20 years.

However, the winning option was to refurbish the old house, because the carbon investment of doing this was just 8 tonnes CO_2e compared with 80 tonnes, and even the highest-specification home could not catch up this advantage over the 100-year period. Once cost was taken into account, refurbishment became dramatically the most practical and attractive option, too.

If this one study is representative, and I suspect that it is, the message for the construction industry is clear. Investment in the very highest levels of energy-efficiency for new homes is, even at its best, an extremely costly way of saving carbon. Investing in improvements to existing homes is a dramatically more cost-effective approach.*

---

*At the time of writing, the UK construction industry is reeling from recession.

100 tonnes *to* 1 million tonnes

Having a child

100 tonnes CO₂e a carbon-conscious child
373 tonnes CO₂e average
2000 tonnes CO₂e high-impact offspring

So if you have two typical children, that's 746 tonnes over their lifetimes.

Unless you will ever contemplate lighting a bushfire (see page 162), the decision to reproduce is probably the biggest carbon choice you will ever make. The more of us there are, the greater the pressure on the world's resources.

I'm not saying you shouldn't have children. If you are someone who believes that God has told you to go and have ten of them, I am not even saying that you are wrong about that. All I'm saying is that according to my sums one of the consequences will be a few thousand tonnes of carbon emissions.

The average lifetime footprint of 373 tonnes CO₂e is based on your child's living to the UK life expectancy of 79 years. I have assumed that during that time he or she will lead a typical UK lifestyle in carbon terms and make average demands on public services. I have also assumed that throughout his or her life the average carbon footprint of a person living in the UK will decrease by 3.9 per cent each year. This is the same annual reduction that is required if the UK is to meet its target of cutting emissions by 80 per cent by 2050 (at which point your child will be roughly half-way through his or her life).*

At the high end of the scale are children who, even after you have done your best to encourage sustainability values, then go on to lead high-carbon lives. At the low end of my scale are children who grow up with carbon priorities embedded in their lifestyle and are serious about reducing emissions where they can.

All my scenarios assume that the child is living in the developed world (the numbers would be much lower in developing countries). For simplicity's sake I have not taken into account the footprint of your child's own offspring.

Deciding whether or not to have children is one thing, but the Optimum Population Trust[1] estimates that 40 per cent of all pregnancies worldwide are unintended and that offering family planning in developing countries saves carbon at a rate of £4 per tonne.

A swimming pool

400 tonnes CO_2e per year

That's the same as 40 people living the 10-tonne lifestyle or just over the expected lifetime footprint of a child born in the UK today.

The figures here are for building public baths with a spa, costing £25 million[2]. They are based on a real study carried out for a pool in a small town in Scotland. The study concluded that the pool in question caused a massive 17 kg CO_2e per visitor, around 30% of which

---

*These calculations do not take account of carbon discount rates. In other words, I have worked from the principle that 1 tonne of carbon emitted today has exactly the same impact as 1 tonne of carbon emitted in the future. This is a broadly reasonable assumption if you take the view that future generations are just as important as our own and that the sensitivity of the planet to each additional tonne of carbon will stay roughly the same throughout your child's life.

could be avoided just through simple improvements in efficiency.

As Figure 10.1 shows, most of the pool's gas was consumed in the process of heating the water. Electricity was used mainly for pumps, air extraction and lighting. Most visitors travelled a fair distance by car to get there, and that accounted for 20% of the footprint. Note that the water itself was barely significant.

Overall, do the high figures here mean we should all stop swimming? There are at least a couple of things to bear in mind before making that leap: the huge social benefits of swimming pools and the fact that an efficient and busy pool in a bigger town could perhaps cut the footprint per swim to one-quarter of the numbers here. Unless you actually believe your local pool should close (in which case there is an argument for avoiding going there), remember that when you swim in it you hardly alter its footprint at all: you just put it to better use. Nonetheless, swimming remains a surprisingly high-carbon way to take exercise.

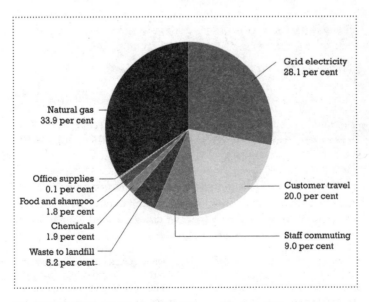

Figure 10.1. The footprint of a swimming pool in a small Scottish town.

A hectare of deforestation

500 tonnes CO$_2$e

That's equivalent to an average car driving 700,000 miles – 28 times around the world.

A hectare is 100 m × 100 m, so there are 100 of them in a square kilometre and about 260 of them in a square mile. Globally we are cutting down or burning about 13 million hectares of rainforest per year. That's about half the land area of the UK. The result is about 9 billion tonnes CO$_2$e or 17 per cent of all man-made emissions.[3]

Most of this total (about 9 million hectares or 6 billion tonnes) involves clearing forest to make way for livestock and other agriculture. One estimate is that 20–25 per cent of rainforest loss is due to cattle grazing, 35–45 per cent to small holdings, 15–20 per cent to intensive agriculture, 10–15 per cent to logging, and perhaps 5 per cent to other causes such as urbanisation, mining, roads and other infrastructure.[4]

Anything that increases the land we need for agriculture drives deforestation. Included in this list are high-meat diets, cut flowers and biofuel crops. In Brazil, where deforestation accounts for 70 per cent of emissions, rates had been falling since 2004 until a spike in beef and soya prices brought on a further increase.

Halting deforestation is potentially one of the easiest climate-change wins, if only we can find the mechanism. Brazil has pledged to cut its levels by 80 per cent over the next decade. That is a big win. The Amazon Fund[5] pays farmers to hang on to their trees. It works out at just £3 per tonne of carbon saved.[6] What a bargain! It is also fantastic for biodiversity. The Norwegian government has pledged US$1 billion to support this. Why doesn't the UK government get into this kind of thing instead of supporting photovoltaic panels (see page 133) at less than one-hundredth of the carbon benefit per pound invested?

A space shuttle flight

At least 4,600 tonnes CO_2e

That's the same as one banana for everyone in the UK or half a sheet of recycled toilet paper for everyone in the whole world.

NASA'S Space Shuttle used to burn through 106 tonnes of hydrogen in its external tank (the big one that would fall off after a couple of minutes and disintegrate before hitting the ground) and 527 tonnes of extra high-energy solid fuel in each of the two booster rockets.[7]

My carbon estimate is conservative for a variety of reasons. I have made the assumption that the process of creating the hydrogen and solid fuel using energy from fossil fuels has been 80 per cent efficient. In other words I've assumed that four-fifths of the energy in the fossil fuel is transferred into the shuttle fuel. That is about as efficient as hydrogen generation ever gets and frankly I would be surprised if energy efficiency was NASA's number one priority. I would be even more surprised if the manufacturing of the solid fuel was that efficient.

Much more significantly, it might have been reasonable to add on a large chunk of footprint of NASA itself. Richard Feynman, the Nobel Prize winning mathematician who helped to investigate the Challenger disaster, describes the Shuttle project as NASA's somewhat unjustifiable *raison d'être* after the lunar landings.[8]

My third major omission is that I haven't included any kind of weighting factor to take account of the high altitude at which the emissions are released. When it is actually in the air, the shuttle burns a different type of fuel from a normal plane and it releases different gases. It won't be the same story as for normal aircraft, since the gases and altitudes are both different. Somewhere along the line the water from burning hydrogen probably causes some contrails, but that's about all I can say. I would love to hear from anyone who knows more about this.

Finally, I haven't bothered to factor in the embodied energy of the shuttle itself. Since each shuttle (apart from Challenger, which crashed after ten trips) was reused around thirty times, I think the emissions of manufacture would turn out to be very small compared to the fuel burn.

Space tourism is not a low carbon option.

A university

72,000 tonnes CO_2e per year

So that's about 8 tonnes for each staff member and student

The figures here are based on a study for Lancaster University. It included just about everything you can think of: gas, electricity, commuting, business travel, everything the university buys, right down to the paper clips. It even included intangibles such as software and banking services. It didn't include everyday food because the university only caters for special events.

As Figure 10.2 shows, gas and electricity between them accounted for 45 per cent of the total. Staff air travel came in at 10 per cent and staff and student car travel came in at about 7 per cent each. Everything else that the university buys made up the remaining quarter of the footprint: IT equipment (5 per cent); buildings maintenance (5 per cent); paper based stuff (1 per cent) and so on.

IT in total accounted for about 12 per cent, with nearly half of that being down to the electricity consumed by computers themselves and a sixth to the power consumed by servers and other computing infrastructure, including the air conditioning to cool it all down. The remaining third was down to the embodied emissions in the equipment itself, with a little bit for services such as internet access and software.

I'd love to be able to write about how Lancaster compares with other

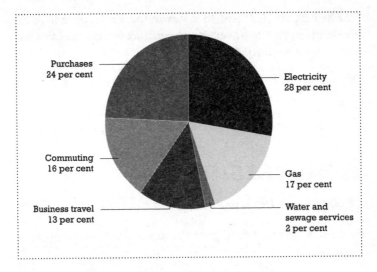

Purchases
24 per cent

Electricity
28 per cent

Commuting
16 per cent

Gas
17 per cent

Business travel
13 per cent

Water and
sewage services
2 per cent

Figure 10.2 The carbon footprint of Lancaster University

universities, but it wouldn't be meaningful, because no two are in the same situation. Lancaster is on a hill in the north of the UK with lots of sixties buildings and a few new ones. Oxford is further south but can be bitterly cold in winter. It isn't on a hill but has to make the best of a load of ancient listed buildings. How could the two be compared? Also, I don't know of another university that has had quite such a comprehensive study carried out. They usually only look at electricity, gas and travel.

Lancaster University's footprint per head is about half the annual footprint of a typical British person. But don't forget that everyone at the university has plenty more in their footprint than the stuff covered by the university. The sums here don't take account of students and staff travelling to and from home, for example.[9]

How to reduce it? I won't go into too much detail here, except to say that at the time of writing, great ideas are coming from all kinds of places. Some are very technical (IT wizardry), some are mind-blowingly simple (change the light bulbs) and some are quirky ideas that bright people have dreamed up out of the blue in odd moments (like bits of hose pipe attached to air conditioning units to improve their efficiency). The biggest challenge is not having the ideas but putting them into action.

Educational establishments account for nearly 2 per cent of the UK's total footprint.

1 million tonnes and beyond

A volcano

1 million tonnes CO$_2$e Mount Etna in a quiet year
42 million tonnes CO$_2$e Mount Pinatubo, Philippines, 1991

If you have been a victim of the rumour, persistent in some circles, that volcanic emissions dwarf those of human activity, now is the time to be liberated. All the world's volcanoes together produce a total of about 300 million tonnes CO$_2$ per year.[1] This is well under 1 per cent of the annual emissions from mankind's activities.

Nonetheless, as the figures above show, each active volcano does have a massive footprint, with a major eruption causing tens of millions of tonnes CO$_2$e. But these numbers are misleading because, alongside their warming effect, volcanic emissions also cause a cooling effect. The ash and sulphur dioxide that they throw up into the stratosphere reflect sunlight away from the Earth. Overall, the Mount Pinatubo eruption of 1991 is thought to have resulted in a net planetary cooling of 0.5°C the following year.

Over time the cooling effect fades faster than the greenhouse effect of the carbon, so the question of whether the warming effect or the cooling effect is greater is not clear cut.

The World Cup

2.8 million tonnes CO$_2$e the 2010 South Africa World Cup

That's 6000 space shuttle fights, three quiet years for Mount Etna, or 20 cheeseburgers for every man, woman and child in the UK.

The headline footprint figure here comes from a study of the 2010 South Africa World Cup and includes players and their entourages travelling around, the construction of the sites, energy used at the stadiums, accommodation, and fans travelling (Figure 11.1).[2]

An estimated 1.2 million spectators will see matches live, so that would be at a massive carbon cost of 2.3 tonnes per viewing. Luckily for the carbon credentials of the World Cup, each of the 64 matches will be viewed on the television by a guesstimated 93 million people world-wide. At 2 hours per match, including intervals, extra times, penalty nailbiters and the bit where they swap shirts at the end, that adds up to a massive 12 billion fan-hours of top-quality entertainment.

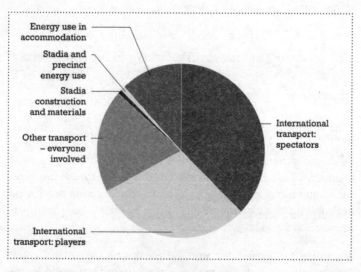

Figure 11.1. A carbon footprint of the World Cup.

If these numbers are correct, the World Cup comes in at 230 g CO_2e per fan-hour of entertainment, though of course the viewers' footprints are boosted by their own televisions. The worst scenario is that you are watching alone on a 42-inch plasma screen (page 30), in which case your TV makes up about half of the footprint of your viewing experience. Even if you watched 24 hours a day, 7 days a week for a whole year (that's a whole year alone without sleep in front of the plasma TV in permanent World Cup ecstasy) you'd only clock up a 4-tonne footprint.[3]

By comparison, a Premiership match, at 820 tonnes, comes out at a tiny 45 g per viewer-hour (excluding all the televisions), with over nine-tenths of the entertainment being exported around the world.

Even better is a kick-around in your local park or street. This is virtually emissions-free, making it one of the best carbon bargains in this book.

The world's data centres

130 million tonnes CO_2e 2010
250–340 million tonnes CO_2e prediction for 2020[4]

The footprint of the world's data centres is currently the same as one-seventh of the UK's footprint, or a quarter of a per cent of the global total.

Data centres are buildings packed top to bottom with computers. These computers store web pages, databases, applications and downloads and generally make the information age possible. As you'd expect, they use lots of electricity (both for powering the machines they contain and for keeping them cool with air-conditioning) and as people consume ever more digital content their already considerable carbon footprint is rising fast.

According to Gartner, the world's data centres currently account for one-quarter of the energy consumed around the world by the information and communication technology sector. That's around two-thirds as much as all the computers and monitors in the world. On current growth trends, however, the power draw of data centres is set to at least double over the coming decade. The precise growth rate will depend on efficiency improvements and changes to the amount of data being stored and processed.

If data centres alone account for 0.25 per cent of the world's total footprint and that figure is set to rise to 0.5 per cent and beyond, then digital data as a whole is looking set to climb to well over 1 per cent of total emissions. Meanwhile 1 per cent is about the proportion of the UK's footprint accounted for by printing and paper-based publishing. The direct comparison is a bit more complex but the point is that digital information may not be lower-carbon than the paper-based world of 20 years ago. Part of the problem is the so called **rebound effect** – the idea that when something (in this case the storing and interrogation of data) becomes cheaper and more carbon-efficient to do, we end up simply doing more of it so that there is no net reduction in cost or impact. Sometimes it is even the reverse.[5] Not only is global data growing incredibly fast, but so is our expectation that we can interrogate it at a moment's notice. Where we might have expected to queue at busy times in a bookshop to enquire about just the contents of its shelves, Amazon now has to meet our expectation, through its data centre capacity, that even in peak times we can search the world's published materials in an instant.

Of course, if we go for digital information without ditching the paper, downloading stuff simply to print it out, we end up with the worst of all carbon worlds.

A bushfire

165 million tonnes CO_2e the Australian bushfires of 2009

That's equivalent to the total yearly footprint of more than 5 million Australians or 50 million Chinese people.

If you were to start one of these deliberately, that one strike of a match would make your footprint thousands of time greater than most people build up over their lifetimes.

My estimate is based on 450,000 hectares (1750 square miles) of forest containing 100 tonnes of carbon per hectare.* To put the number in perspective, the most recent estimate of Australia's annual footprint was 529 million tonnes CO_2e, so the fires added nearly one-third.**

Emissions from bushfires vary from year to year. In 1997–98 they are thought to have been around 2.1 billion tonnes.[6] Although some vegetation regrows, they are almost certainly a nasty example of a climate change positive feedback loop.

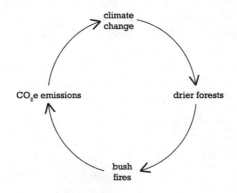

In theory, regrowth will absorb the CO_2 from the air in time, thus making the fire carbon neutral in the long term. However, it is looking increasingly likely that permanent changes in terrain are taking place. Furthermore, bush fires are also a major source of black carbon.

---

*The figures I used are at the low end of the estimates given. The 2009 Australian bushfires covered at least 450,000 hectares (Wikipedia, 2009). Asa Wahlquist, in *The Australian* (13 February 2009), quotes Mark Adams from the University of Sydney in reporting that the area affected contained over 100 tonnes of carbon per hectare. Carbon forms 3.67 times its weight in CO_2.

**This figure does not include an emissions weighting multiplier to take account of the additional impact of high-altitude aviation emissions (A report by Ecofys for the UK's Department for Environment, Food and Rural Affairs in 2007 'Factors Underpinning Future Action. 2007 Update')

A country

Around 1.5 million tonnes CO$_2$e per year Malawi
4 million tonnes CO$_2$e per year Iceland
610 million tonnes CO$_2$e per year Australia
810 million tonnes CO$_2$e per year France
862 million tonnes CO$_2$e per year the UK
890 million tonnes CO$_2$e per year Canada
980 million tonnes CO$_2$e per year Brazil
1350 million tonnes CO$_2$e per year Gemany
1400 million tonnes CO$_2$e per year India
1940 million tonnes CO$_2$e per year Russian Federation
4300 million tonnes CO$_2$e per year China in 2005
8250 million tonnes CO$_2$e per year the US

The estimates here are of the footprints of national consumption in 2005. They include the footprint of goods imported from overseas but exclude goods produced in each country for export.

Let's start off by looking at the footprint of the UK. As we saw earlier (see page 7), each person in the country is responsible for around 15 tonnes CO$_2$ per year. Because the UK imports more goods than it exports, the total figure for the country – 862 million tonnes – is higher than the 700 million or so tonnes that usually gets reported for UK emissions.[7]

Taking the UK as an example of a fairly typical European country, let's see how all those emissions break down (Figure 11.2).

Domestic energy, which often dominates the media coverage of carbon footprints, makes up 22 per cent of the total, consisting of household fuel at 13 per cent and electricity at 9 per cent. For most people the fuel is gas, for which 84 per cent of the emissions happen in the home itself and the rest are caused during gas extraction and distribution.

Cars come in at 15 per cent of the total when you add together the 10 per cent caused by their fuel (extraction, distribution and use) and the 5 per cent caused by their manufacture and maintenance.

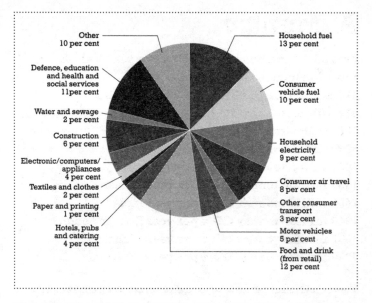

Figure 11.2. A breakdown of the UK's carbon footprint, including imports but excluding exports.

As a rule of thumb, the exhaust-pipe emissions are about half the total footprint of driving (see page 65). Note that this slice of the pie doesn't include commercial vehicles, so the whole of road transport is a good deal more.

Food and drink, often underestimated, comes in at 12 per cent just for those groceries bought at shops. If we include all the food and drink served by hotels, pubs, cafes, schools, hospitals and so on, we'd get to about 17 per cent. If we also added in the emissions from cooking at home, travelling to the shops and the emissions from food waste sent to landfill, the total footprint of the stuff that goes into people's mouths comes to about 20 per cent of the UK's footprint. It's roughly the same percentage for the world as a whole. All these numbers are without considering the impact of food demand on deforestation, which would take the UK total to around 30 per cent.[8]

Air travel for private purposes is a staggering 8 per cent[9] of the total. If you include business travel and air freight as well, flying comes in

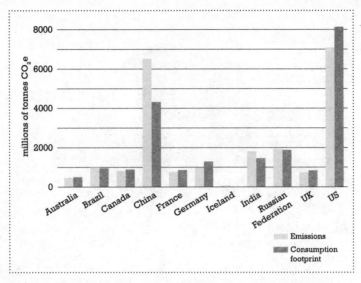

Figure 11.3. National emissions and consumption footprints in 2005.

at around 12 per cent of the UK's footprint – much higher than the figure usually quoted. British people probably fly a bit more than other Europeans because they live on an island and because there is a lot of sea to our west, but this is still a remarkable statistic – especially when you consider that air travel is the fastest-growing major emissions source in the country.

In the pie chart, **construction**, at 6 per cent of the total, includes domestic repairs, new houses and all new commercial construction work. And the production of **electrical goods** – that is, household computers and appliances – comes in at 4 per cent, almost half as much as the electricity they consume in use.

Public administration, defence, education, health care and social services cause a significant 11 per cent of emissions. A common misconception is that there is nothing we can do about this as individuals. To cut your share of these emissions, how about avoiding crime, encouraging schools, universities and businesses to manage their carbon, staying as healthy as possible, and voting with climate change in mind?

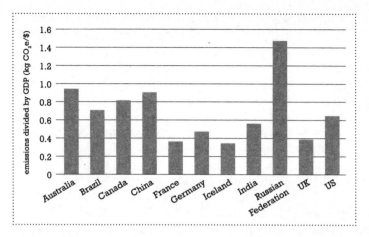

Figure 11.4. Emissions per dollar of GDP for a selection of nations.[10]

Other factors. This is my catch-all category. It contains a jumble of things, including some that might never occur to you as incurring a footprint at all. In here are bikes, brooms, Lego bricks, lipstick, legal fees, phone calls, footballs, tables, toiletries, travel insurance, jewellery and too many other things to list individually.

So that's the UK. How about the rest of the world? Figure 11.3 picks out a few key countries and gives their official carbon emissions (that is, the quantity of greenhouse gases released within their borders) and my estimate of their true consumption footprints (that is, with exports subtracted, and imports and international travel and transport added in). Note that the numbers shown are from the year 2005. Things change fast, and China is now widely thought to lead the US in terms of emissions, probably having drawn level in about 2006, even though the footprint of Chinese consumption is still just over half that of the US.

Even when you factor in imports and exports, however, it isn't necessarily hugely meaningful to think of the total emissions of a country. China may have overtaken the US as the biggest emitter, for example, but it has *far* more people. Hence emissions per person is usually a more meaningful way to compare one nation with another – as discussed on page 197.

Yet another way of looking at a country's footprint is in terms of emissions per unit of GDP (gross domestic product); see Figure 11.4. This is a measure of 'carbon efficiency' or 'carbon intensity' – a nation's footprint relative to its economic activity. Countries with inefficient factories, and which get their electricity from dirty coal-fired power stations, rate worse on this scale. Hot countries can sometimes achieve a better rating because they don't have to spend so much on keeping warm (provided that people aren't rich enough to afford air-conditioning).

Through this GDP lens, Russia comes out worst, because of its coal-fired power stations, inefficient factories and cold climate, with coal-dependent China and Australia following behind. Western Europe has relatively efficient factories and cleaner electricity, so countries in this region come out well – especially nuclear-powered France and renewable-powered Iceland. The US comes in somewhere in the middle of the carbon efficiency stakes.

In the UK we talk about becoming more carbon efficient. It is clear, however, that we are a very long way from being able to grow our economy without increasing our consumption footprint. Tim Jackson's recently published *Prosperity Without Growth*[11] is both the most rigorous and the most accessible articulation of this uncomfortable reality that I have seen.

Ultimately, there's no avoiding the fact that a country's emissions are strongly linked to its wealth. It's hard to be rich and have a low carbon footprint (see Spending £1, page 59). Malawi is just one example of a country whose poverty ensures a low footprint. Its 14 million people have a footprint of around 100 kg each per year.

I've looked at typical footprints by country, but this doesn't always give the full picture. Sometimes the most significant differences occur *within* countries. In China, for example, hundreds of millions of people live very low-carbon lives, whereas the emerging middle class, with Western lifestyles in a less energy-efficient economy, probably have carbon footprints to dwarf those of the Australians.

A war

690 million tonnes CO$_2$e a 'limited' nuclear exchange of
50 fifteen-kilotonne[12] warheads
250–600 million tonnes CO$_2$e Iraq, 2003–09

**The Iraq war has so far probably racked up a
carbon footprint roughly equivalent to the whole
of the UK economy for between 3 and 8 months.**

The direct human costs of wars are so great that it might seem flip-
pant to think about their climate change costs. But war unfortunately
plays a big role in global society, so this book wouldn't be what it
says on the cover without giving it a mention. Moreover, it's worth
bearing in mind that even just the emissions of a war could ultimately have serious human impacts somewhere in the world.

In what was perhaps the only academic estimate of the carbon footprint of an atomic war it was concluded that even a 'small nuclear
exchange' of just 50 fifteen-kilotonne warheads would cause 690
million tonnes of CO$_2$ emissions through the burning of cities.[13]
The same report also estimated that the exchange would also release
313 million tonnes of soot into the atmosphere, which would have a
cooling effect and would therefore counter the warming for the first
few years after the explosions.

But a war doesn't need to be nuclear to have a huge carbon footprint.
At the time of writing the financial cost of the US military operation
in Iraq since 2003 has been estimated at $1.3 trillion so far, with
a further $600 billion anticipated for the lifetime healthcare costs
of injured troops.[14] We can use the input–output model to give a
very crude estimate of the footprint of the US operation, of 160–500
million tonnes CO$_2$e for the military activities and perhaps a further
80 million tonnes for the healthcare of troops.[15] This excludes the
actual emissions from combat itself. Add on a few per cent to both
numbers to include the coalition forces. Also add perhaps another 1
per cent for the footprint of the much more poorly resourced insur-
gency. Overall we might be looking at 250–600 million tonnes –

roughly equivalent to everyone in the UK flying to Hong Kong and back between one and three times.

The war-and-carbon discussion starts to get distinctly uncomfortable (and methodologically just about impossible) at the point where we start factoring in the indirect emissions impact caused by the death toll and indeed the broader economic impacts of the war. In the nuclear example, the report in question estimates 17 million deaths – equivalent to around one-quarter of the UK population. Looked at in the starkest and simplest possible terms, if each of these people had a typical UK footprint, then the carbon saving of their ceasing to exist might make up for the direct emissions from the war in just a few years. In other words, mass annihilation turns out to be an effective way of curbing emissions – though of course it also defeats the object.

Black carbon

7–15 billion tonnes CO_2e per year

That is 15–30 per cent on top of the figure I normally quote for global man-made emissions.

How can this have slipped off the radar for so long?* As Dennis Clare of the World Watch Institute put it, 'Black carbon, a component of soot, is a potent climate forcing aerosol and may be the world's second-leading cause of global warming after CO_2.'[16]

My low figure for black carbon's global warming impact – 7 billion tonnes of carbon equivalent – came from the Intergovernmental Panel on Climate Change 2007 report. The higher figure of 15 billion tonnes came from more recent studies.[17]

Black carbon warms the world in two ways. In the atmosphere it con-

---

*I inserted this entry late in the day. The implications of black carbon are not systematically integrated throughout the book. I could apologise, but then, it's just an example of how we are all having to learn on the go.

tributes to the greenhouse effect. Down on the ground, it turns snow and ice murky and in so doing makes it absorb more of the Sun's heat. It is thought to be a major contributor to the global reduction in ice cover, especially in the Northern Hemisphere.

Black carbon is caused by incomplete combustion; 42 per cent comes from outdoor fires of one kind or another, and one-quarter comes from the burning of wood, coal, dung, peat and any other organic stuff in homes. A further quarter comes from transport (mainly diesel), and about 10 per cent comes from coal-fired power stations.

The good news about black carbon is that it lasts only a few days in the atmosphere. In other words, if we can reduce the amount we create, the benefit will be instant. Hence some experts think that reducing black carbon pollution should be a number one priority in tackling global warming. Easy wins can be made by using particulate filters on diesel engines and swapping inefficient open fires for super-efficient stoves.

The world

50 billion tonnes CO_2e per year

In 2007, the IPCC estimated global greenhouse gas emissions at 49 gigatonnes (that is, 49 billion tonnes) CO_2e a year and rising. That doesn't include a multiplier to take account of the extra warming effect of emissions from planes, which takes the total to 50 gigatonnes CO_2e. Figure 11.5 shows the emissions broken down to constituent greenhouse gases, looked at in terms of the impact over a 100-year period.

Around half the methane comes from agriculture (especially livestock at 5 per cent of global emissions, but also rice cultivation at 1.5 per cent and other farming). The rest of the methane comes mainly from the extraction and processing of coal, gas and oil, from landfill (2 per cent of the global total), from the treatment of used water and from other wastes.

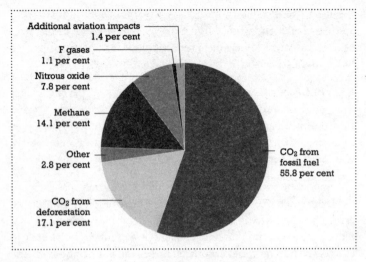

Figure 11.5. The breakdown of our 50 billion tonnes CO_2e greenhouse gas.[18]

Nitrous oxide results mainly from the spreading of nitrogen fertilizer and manure, although there are also contributions from fuel combustion, industrial processes and waste treatment. The F gases, at 1.1 per cent, result mainly from refrigeration and air-conditioning (there are some good technologies coming through to deal with this: see Refrigeration, page 53). Aviation comes to 3 per cent of the total once you factor in the effect of altitude (in the pie chart the 1.4 per cent sliver is just the additional bit to take account of high-altitude effects).

Although there is a convention to look at the impact of different gases over a 100-year timespan (that is, up until 2110), this is in fact somewhat arbitrary. Many people believe that climate change will bite much sooner than that, so there may be a case for considering the impact over shorter timescales as well. This changes the relative impact of the different gases: those that are powerful but short-lived become more important relative to the weak but long-lasting CO_2. If you were to look at a 50-year timescale, for example, the non-CO_2 emissions caused by agriculture, refrigeration and air-conditioning would immediately become roughly twice as serious.

Figure 11.6. The world according to greenhouse gas emissions (as of 2000).[19]

To understand *where* in the world all these emissions come from, it's time for some squidgy maps. The first, Figure 11.6, shows the size of countries in proportion to their emissions. Note that these are just the emissions that physically rise out of each country rather than the consumption footprint, so the emissions from a factory in China that makes washing machines for people elsewhere in the world is shown as belonging to China, rather than to the people who buy them. The

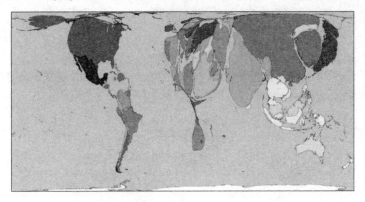

Figure 11.7. Increases in global greenhouse gas emissions, 1980–2000.[20]

US, China and Europe dominate the map. More recently, China has overtaken the US as the biggest national emitter.

Figure 11.7, which focuses on emissions growth, shows how things have been changing. In this view, China, India and the US dominate, while most of Africa has been strangled to nothing. A few European countries have vanished, too, showing that their emissions were static or declining.

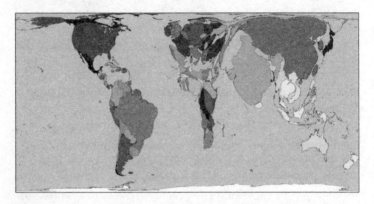

Figure 11.8. Global emissions of methane and nitrous oxide in 2000, in CO_2e.[21]

Figure 11.8 shows just methane and nitrous oxide, and in doing so it emphasises agricultural emissions. In this view China, India and South America dominate. North America, although hardly looking lean, is at its smallest. This is by far the most imposing view of Africa out of the three, although it still looks emaciated. A bit more fertiliser application in Africa could actually be helpful.

Burning the world's fossil fuel reserves

2.5 trillion tonnes CO_2e

That's 50 years of current global emissions.

The exact figure depends on just how big you think our reserves are. The numbers in Table 11.1 are based on 'proven' reserves, as of 2006, but nobody really knows for sure how much is down there. Because fossil fuels account for only a little over half of total global emissions, even this conservative estimate of our reserves means that we have enough fuel left to allow us to keep on belching out carbon at our current rate until roughly the end of the century.

In other words, regardless of the precise amount of fossil fuel left in the ground, it's clear that there's more than enough to push us into climate meltdown if we were to burn it all – and judging by our current mindset it does look as though we are on course to do just that. For this *not* to happen, we will need to achieve a situation in which although there is fuel in the ground waiting to be extracted and burned, the countries and businesses that own the rights to it are simply content to leave it there for all time.

For me this is an uncomfortable perspective because it is hard to

Fuel	Billions of tonnes oil equivalent	Billions of tonnes CO_2e from burning
Coal	463	1470
Oil	165	530
Gas	163	520
Total	791	2520

Table 11.1. Proven fossil fuel reserves[22]

imagine Russia, China, the US, Saudi Arabia, Exxon, Shell, BP or anyone else simply being happy to leave their valuable assets down there in the ground. One solution, albeit a long way off, might be to devalue those assets by making renewable energy even more abundant.

More about food

Food deserves a chapter of its own because in terms of carbon footprints it is such an important but poorly understood area. The various food entries early in this book have covered many of the key points, but this chapter briefly pulls together all the main issues to give a sense of the food sector overall.

As we saw earlier, the food we buy adds up to around 20 per cent of our carbon footprint.[1] In the UK that's 170 million tonnes CO_2e per year – nearly as much as household fuel and electricity put together. If you factor the effect of deforestation, the footprint of our food goes up again to a staggering 30% of the UK total. Interestingly, food is also a very expensive part of our footprint. If you want to trash the planet, buying the wrong food or wasting what you buy is a far more expensive way of going about it than leaving the lights on or turning up the thermostat.

How the footprint of food breaks down

Figure 12.1 shows an estimated breakdown of the footprint of food at the point when it leaves a UK supermarket. This is just a best estimate, based on work I've been doing for about 3 years for the Booths supermarkets chain. As the chart shows, two-thirds of the impact is on the farm. Transport is a big deal for some products but not most.

The supermarket's own operations make up about one-ninth of the total picture.

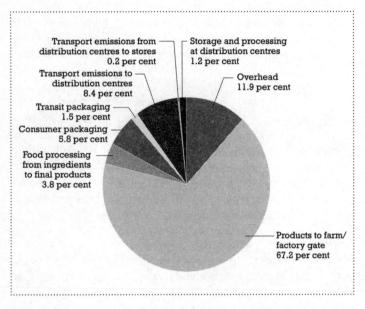

Figure 12.1. Total footprint of Booths products and supply chains: 226,000 tonnes CO_2e.

Farms

Whereas CO_2 is the dominant greenhouse gas overall, it accounts for only 11 per cent of the UK's agricultural emissions.[2] The rest is nitrous oxide (53 per cent) and methane (36 per cent). Nitrous oxide is 296 times more potent per kilo than CO_2 as a climate change gas and on farms it results mainly from the use of fertiliser, but also from cattle pee, especially if there is excessive protein in their diet, and from the burning of biomass and fuel.[3] Methane, which is 25 times more potent than CO_2, is mainly emitted by cows and sheep belching. Some is also emitted from silage. The CO_2 comes from machinery but also from the heating of greenhouses to grow crops out of season or in countries that just don't have the right climate.

Transport

The first thing to say about transport emissions is that for all the talk that we hear about food miles, they are not the most pivotal thing to think about. At Booths, over one-quarter of the transport footprint comes from the very small amount of air freight in their supply chains – typically used for expensive items that perish quickly. Conversely, most of their food miles are by ship, but because ships can carry food around the world around 100 times more efficiently than planes, they account for less than 1 per cent of Booths' total footprint. The message here is that it is OK to eat apples, oranges, bananas or whatever you like from anywhere in the world, as long as it has not been on a plane. Road miles are roughly as carbon intensive as air miles, but the distances involved tend not to be too bad. Booths is a regional supermarket with just one warehouse, so UK delivery is more efficient than most, and they have been working hard on further improvements.

Meat and dairy

Food from animals turns out to be more carbon intensive (remember, this is my shorthand for *greenhouse gas intensive*) than food from plants, simply because animals are inefficient devices for producing food. They eat plants and then spend their lives wasting most of the energy from them on things such as walking around and keeping warm. It is a far more efficient process for humans to eat plants directly, so that all the plant energy can go directly to us. Beef and lamb are doubly high in carbon because of all those belching ruminants. Chicken is a bit better because, to put it bluntly, they don't live as long, so they don't get so much opportunity to waste the energy in their feed.

Dairy has all the same problems of ruminant meat production, so there is little point in switching from beef to cheese. A kilo of cheese comes in at around 13 kg CO_2e, compared with around 17 kg for beef. Milk comes in at around 1.3 kg per litre.

Hot houses

Protected crops can be just as impactful as air-freighted foods. It takes a huge amount of energy to keep a greenhouse warm enough to grow tomatoes during the winter (see page 86).

Packaging

This topic needs to be kept in perspective. It's only about 6 per cent of what you should be considering as you shop. And at its best it serves a purpose, helping food to stay fresh and letting you know what you are buying. Indeed, a simple bag can dramatically improve the shelf life of some fresh foods.

At Booths we found that no single material dominated the packaging footprint, and there were some surprises.

- ■ Paper and card are often more carbon intensive than plastic packaging, mainly because making paper is so energy intensive but also because it emits methane if it ends up in landfill.

- ■ Plastic is environmentally nasty as either landfill or litter because it hangs around for so long. However, it is typically not quite as energy intensive to produce as card packaging and has the advantage, from a purely carbon perspective, that when you put it in landfill, you are just sending those hydrocarbons back into the ground where they came from for long-term storage. In the days when supermarkets routinely gave out disposable plastic bags, they accounted for around one-thousandth of the footprint of a typical shopping trip. Biodegradable plastic can be a well-intentioned nightmare, clogging up recycling processes, with potential to ruin a whole batch. In landfill it rots, emitting methane.

- ■ Glass is energy intensive to make (or recycle) and its weight adds to the transport footprint. Cans of beer are better than bottles, as are cartons or boxes of wine. Incidentally, bottles are absolutely no better for storing wine than the more climate-friendly alternatives.

- Steel and aluminium are carbon-intensive stuff, but you don't need a great weight of them and they're easy to recycle. It takes only about one-tenth of the energy to recycle aluminium compared with extracting it from ore in the ground.

Food waste

In the developed world we are thought to waste about one-quarter of the edible food we buy.[4] This figure depends partly on your definition of what was edible in the first place. Do you think of the potato skin as just packaging or do you think of it as the tastiest and most nutritious bit? Whatever your definition, a huge and expensive proportion of our food gets left on plates, is allowed to go off in the fridge, isn't scraped out of the pan properly or isn't picked off the carcass. It is slightly better to compost waste food than to throw it into landfill, but it doesn't get you away from the main issue that the carbon footprint of that food has been needlessly incurred.

Refrigeration

Fridges use electricity, and it takes energy to make them in the first place. On top of that is the problem that traditionally they have relied on the use of refrigerant gases that have a global warming potential several thousand times that of CO_2. This stuff tends to leak out of large commercial fridges, which need topping up regularly. At Booths, this leakage from within the stores and warehouse accounted for around 3 per cent of the total footprint. And refrigeration accounted for about half of all electricity usage in stores. When all considerations are taken into account, refrigeration probably accounts for around 6 per cent of the footprint of supermarket food.

There are huge strides being made in cooling technologies. These include the use of other gases with dramatically lower global warming potential,[5] the re-use of spare heat to warm the stores, and the use of underground cooling pipes. Booths is starting to employ CO_2-based refrigeration systems (thereby almost eliminating the climate change impact of gas leaks) and expects to have replaced almost all its fridges with these within a decade. The company is also re-using the heat in their newest stores. Thanks to these kinds of approaches, we can expect the footprint of commercial refrigeration to fall dramatically. In the meantime, do not let it put you off your fresh, chilled produce.

Low-carbon food tips

The following is a quick summary of the various steps you can take to reduce the carbon footprint of your diet – and the type of saving you can expect.

Eat what you buy. Ask people how much they would like before you serve them. Eat the skins. Clean the plates, pick the carcass. Save the leftovers. Check what needs eating when you plan your menus. Keep vegetables in the fridge if you can. Rotate the contents of your cupboards so the old stuff is at the front. Eradicating waste is worth a 25 per cent saving for the average shopper.[6]

Reduce meat and dairy. I'm not saying go vegan any more than I'd say never drive. But there is no dodging the fact that meat and dairy is a key area. By reducing our consumption of these food types, many of us will live a bit longer and save money as well as reducing our emissions. The vegetarians and vegans I know don't consider it a hardship. Sensible reductions in meat and dairy without needing to go vegetarian are probably worth another 25 per cent saving on a typical UK diet.

Go seasonal, avoiding hothouses and air freight. Local seasonal produce is best of all, but shipping is fine. As a guide, if something has a short shelf life and isn't in season where you live, it will probably have had to go in a hothouse or on a plane. In the UK in January, examples are lettuce, asparagus, tomatoes, strawberries and most cut flowers. Apples, oranges and bananas, by contrast, almost always go on boats. Adopting this tip religiously can probably deliver a 10 per cent saving on a typical UK diet.

Avoid low-yield varieties. Cherry tomatoes and baby corn are classic examples. Estimated saving: 3 per cent.

Avoid excessive packaging. Some packaging serves a valid purpose in keeping food fresh. But a metal dish inside plastic trays inside a plastic bag within a cardboard box is probably excessive. Worth around 3–5 per cent.

Recycle your packaging. Worth 2–3 per cent.

Help the shop reduce waste. Always take from the front of the shelf so that the stock can be rotated. Handle food with care. Buy the reduced-price items when you can, but don't hang around waiting for them to be reduced. Worth perhaps a 1 per cent saving.

Buy misshapen fruit and vegetables. Stimulate demand for the huge quantities of produce that get thrown away just because of their shape. The savings are hard to quantify, but perhaps 1 per cent.

Lower-carbon cooking. Use a pan lid whenever you can. Remember that water boils at the same temperature however much heat you apply, so for cooking food, a gentle boil is just as fast as a furious one. Turn out the gas when you are not using it. Use a microwave when appropriate. Perhaps a 5 per cent saving.

Incredible! The savings here add up to about 75 per cent. Sadly the maths doesn't work out quite like that because some of these points overlap. If you do them all it works out more like a 60 per cent saving – still a remarkable amount.

A guide to seasonal food

Here is a quick guide to what is in season at different times of year in the UK.[7] Similar lists can be found online for other countries.

Vegetables and salads

January: beetroot, broccoli, broccoli (white and purple), brussels sprouts, cabbage, carrots, celeriac, chicory, Jerusalem artichoke, kale, leek, parsnip, potatoes, purple sprouting broccoli, shallots, spinach, squash, swede, turnips.

February: beetroot, broccoli, broccoli (white and purple), brussels sprouts, cabbage, carrots, cauliflower, celeriac, chard, chicory, Japanese mustard spinach (komatsuma), kale, kohlrabi, leek, parsnip, potatoes, purple sprouting broccoli, spinach, squash.

March: beetroot, broccoli, broccoli (white and purple), brussels sprouts, cabbage, carrots, cauliflower, chicory, claytonia, Japanese mustard spinach (komatsuma), kale, leek, parsnip, potatoes, purple sprouting broccoli, radishes, salad onions, spinach, spring greens, watercress.

April: asparagus, broccoli, broccoli (calabrese), broccoli (white and purple), brussels sprouts, cabbage, carrots, cauliflower, chicory, Japanese mustard spinach (komatsuma), kale, leek, potatoes, purple sprouting broccoli, radishes, salad onions, spinach, spring greens, watercress.

May: asparagus, broccoli, broccoli (white and purple), cabbage, carrots, cauliflower, chicory, chopsuey greens, land cress, lettuce, mushrooms, new carrots, new potatoes, potatoes, radishes, rocket, salad onions, spinach, spring greens, watercress.

June: asparagus, aubergine, broad beans, broccoli, cabbage, carrots, cauliflower, celery, chopsuey greens, courgette, cucumber, French beans, kohlrabi, land cress, lettuce, mangetout, mushrooms, onion, peas, peas (shell), peas (sugar snap), peppers, potatoes, radishes, rocket, runner beans, salad onions, spinach, spring greens, spring greens (undeveloped spring cabbages), squash, texel greens, tomatoes (indoor), turnips, watercress.

July: aubergine, beetroot, broad beans, broccoli, cabbage, carrots, cauliflower, celery, chopsuey greens, courgette, cucumber, Florence fennel, French beans, globe artichoke, kohlrabi, land cress, lettuce, marrow, mushrooms, okra, onion, peas, peas (shell), peas (sugar snap), peppers, potatoes (new), pumpkins, radishes, rocket, runner beans, salad onions, shallots, spinach, spring greens, squash, texel greens, tomatoes, turnips, watercress.

August: aubergine, beetroot, broad beans, broccoli, cabbage, carrots, cauliflower, celery, Chinese celery, chopsuey greens, courgette, cucumber, Florence fennel, French beans, globe artichoke, kohlrabi, land cress, lettuce, marrow, mushrooms, okra, onion, peas, peppers, potatoes, pumpkins, radishes, rocket, runner beans, salad onions, shallots, spinach, spring greens, squash, sweetcorn, texel greens, tomatoes, turnips, watercress.

September: aubergine, beetroot, broccoli, broccoli (calabrese), brussels sprouts, cabbage, carrots, cauliflower, celeriac, celery, chicory, Chinese celery, chopsuey greens, cucumber, endine, Florence fennel, French beans, globe artichoke, Japanese mustard spinach (komatsuna), kale, kohlrabi, leeks, lettuce, marrow, mushrooms, okra, onion, parsnip, peas, peppers, potatoes, pumpkins, radicchio, rocket, runner beans, salad onions, shallots, soya bean, spinach, spring greens, squash, swede, sweetcorn, tomatoes, turnips.

October: beetroot, broccoli, broccoli (calabrese), brussels sprouts, cabbage, carrots, cauliflower, celeriac, celery, chicory, Chinese celery, chopsuey greens, claytonia, corn salad, courgette, endine, Florence fennel, French beans, Japanese mustard spinach (komatsuna), Jerusalem artichoke, kale, kohlrabi, leeks, lettuce, marrow, mushrooms, okra, onion, parsnip, potatoes, pumpkins, radicchio, runner beans, salad onions, shallots, soya bean, spinach, spring greens, squash, swede, sweetcorn, tomatoes, turnips, watercress.

November: beetroot, broccoli, broccoli (calabrese), brussels sprouts, cabbage, carrots, cauliflower, celeriac, celery, chicory, corn salad, endive, Japanese mustard spinach (komatsuna), Japanese turnip, Jerusalem artichoke, kale, kohlrabi, leeks, onion, parsnip, potatoes, pumpkins, salad onions, soya bean, spinach, squash, swede, turnips.

December: beetroot, broccoli, brussels sprouts, cabbage, carrots, cauliflower, celeriac, celery, chicory, corn salad, Japanese mustard spinach (komatsuna), Japanese turnip, Jerusalem artichoke, kale, leeks, onion, parsnip, potatoes, spinach, spring greens, squash, swede, turnips.

Fruit

All year round: Apples, pears, oranges, citrus fruit and, of course, bananas, pineapples and mangoes can be low carbon even if they have come by boat from the other side of the world or stored. The following are also likely to be in season.

January: All-year-round fruit – as above.

February: early rhubarb.

March: rhubarb (forced).

April: rhubarb.

May: gooseberries, rhubarb, strawberries.

June: cherries, gooseberries, raspberries, redcurrants, strawberries, rhubarb.

July: blackcurrant, blueberries, cherries, gooseberries, loganberries, peaches, raspberries, redcurrants, rhubarb, strawberries.

August: blackberries, blueberries, cherries, gooseberries, greengages, nectarines, peaches, loganberries, raspberries, strawberries.

September: apples, blackberries, blueberries, damsons, grapes, melons, nectarines, pears, peaches, plums, raspberries, strawberries.

October: apples, elderberries, grapes, pears.

November: apples, cranberries, pears.

December: apples, pears.

Some more information

Some assumptions revisited

I started out with three assumptions:

- Climate change is a big deal.
- It is man made.
- We can do something about it.

This book isn't really about those assumptions, but this section is for anyone who is still unsure. The human capacity for collective denial is an amazing phenomenon to watch. If that is where you are right now, I'm not too hopeful that I can shift you.

Is climate change a man-made big deal?

At the end of the day we all have to make up our own minds. I can't go over the scientific arguments in detail here, and even if I did I'd just be one more voice for you to sift through. But I will briefly go through how I came to make up my mind.

None of us really knows for sure what climate change is going to mean for us in the coming decades. The science is hideously complex and uncertain. The media still report a full spectrum of arguments. It's a confusing picture for the layman. What basis can we have for knowing whether a news article, a TV programme or a book is credible?

A key question in this context is *how can we work out whom to trust?* I meet plenty of people who have understandably given up trusting anyone over climate change. But it is possible to do a lot better than that. This is how I make up my own mind about a report or a piece of research:

1. I look at the argument itself and see if the logic makes sense at face value.

2. I look at the competence of the source.

3. I look at the resources and information that it had at its disposal.

4. Critically, I try to understand the motivations – political, financial and psychological. How strong was the dedication to truth? Who funded it and what did those funders want? Who wanted what from their careers, and what influence might this have had? What was the psychological readiness of the source to accept and report on different findings that might emerge?

These are the questions I have been asking about sceptics' arguments. They can sometimes pass the first test but every single one of them fails at least one of the final three.

A few years back, just before I reoriented my working life towards addressing climate change, I thought I'd better double check that the whole thing wasn't a storm in a teacup. I didn't want to go to a whole lot of trouble for nothing. I knew my family was going to have to put up with my hardly earning anything for a year or two while I learned a new trade.

A good friend of mine had raved about Bjørn Lomborg's book, *The Skeptical Environmentalist*. 'Mike,' he said, 'I've read this book and it's rearranged my thinking.' It's a thick and persuasively written tome with some 2000 academic references. It makes the claim that we can all afford to chill out about climate change and we would do better to invest the money elsewhere. Lomborg further asserts that the climate-change worriers are psychologically wedded to a doom-and-gloom position on life. To me, that last point hit a nerve. It was an important challenge to address. I thought, 'Perhaps he's right! Maybe I should ask myself if this applies to me?' I didn't want the experience of realising in years to come that the only reason I've done all this stuff about climate change is because of some unhealthy personal hang-up. At the very least I felt that the mainstream scientific community should have a blisteringly clear response to Lomborg, and it was disquieting that I couldn't readily find one.

I sat down to spend about a week with Lomborg's work. I picked into some of his arguments in detail and before long found that even from my distant position I could see several clear misrepresentations of science. Then I found that his book had never been peer-reviewed. Then I started uncovering websites that detailed his errors literally in their hundreds, along with roasting dismissals of his arguments from scientists, statisticians and economists alike. After that I started to read about Lomborg's close shaves with the Danish Commission for Scientific Dishonesty. In the end it was abundantly clear to me that the whole thing was a sham. I came to a clear view, but it took detailed consideration of his work; far more than can be expected of the average man on the street. Lomborg passed the first and third of my tests but failed the second and fourth. To this day Lomborg carries on, and has a following. It is incredibly unhelpful for the world. I don't know any scientists who have any time for his position at all, although some commentators treat his work with unwarranted respect in the misguided name of 'balance' or perhaps just to be polite.

In the name of open-mindedness I've looked in detail at several other 'sceptics' and had a similar experience.[1]

So much for the sceptics. Let's look at the mainstream scientific community. The UN's Intergovernmental Panel on Climate Change consists of around 2500 scientists. The sceptics point out that there may be potential for group-think and mass hysteria. These are warnings that should be taken seriously. Furthermore, there have been occasional errors in the IPCC's work, and even the hint of the odd deliberate misrepresentation. However, the standard of integrity that is demanded of the climate-change believers is on a different plane altogether from that demanded of the sceptics. As I write, some scientists at the University of East Anglia are in headline-hitting trouble for 'sexing up' their work in a way that the some of the sceptics would consider quite normal.

It's worth bearing in mind that it would also be possible to criticise the IPCC for its caution. Does it offer a sufficient platform for the airing of discomfort about poorly understood scientific risks? Does the level of deliberation and the need for consensus among such a wide community, some members of which have clearly been under political pressure to play things down, result in an undercooked estimation of the risks? We can't know for sure. We do know that the extent of scientific consensus is almost unanimous in affirming the first two of my assumptions.

Finally I want to note a trend that I have also picked up on among the people I know. The more scientifically minded they are and the more they have thought about the issues, the more worried they tend to be that even though we *might* almost all be fine, it is also just as likely that we'll end up frying in our billions. I talk to a lot of academics, mainly physical scientists and social scientists. In the last few weeks I've started conducting my own informal opinion poll by asking any senior academic that I meet to estimate the percentage of people in their department who think that 'climate change is a big deal and is man made'. So far I have yet to have anyone give me a figure under 99%. It is an amazing phenomenon that the academic community, those with the most realistic and mature understanding of how the academic process works and of how scientific knowledge evolves, are so clear about my first two assumptions whilst the wider public remains so obstinately doubtful.

Can we do something about it?

People ask me sometimes why they should bother when, even if everyone in their country cut the carbon, it would make such a small impact on world emissions. Sometimes I hear businesspeople trying out the argument that their hands are tied until governments act or until their end consumers care more. Governments say they can't move ahead of popular opinion. I hear Chinese people saying that the developed world started it and is more carbon hungry so they should start the cuts, whereas in the UK I hear people saying we're just a pinprick in comparison with the US or the emerging Chinese middle classes.

The UN climate negotiations in Copenhagen and elsewhere have surely taught us that it isn't enough to hope that world leaders will sort things out on their own. So the question is: Where does leadership come from? My answer is that it can come from anywhere and we need it to come from everywhere at once. If the Chinese middle class wants a Western lifestyle, then Western lifestyles had better become lower carbon. Who can start that off? Anyone can. Anyone who finds a way of enjoying life more for less carbon is setting a standard for others. Anyone who chooses a lower-carbon food is helping the supermarkets to emphasise that product. Any supermarket that improves and promotes its lower-carbon range is helping its customers to enjoy low-carbon food. All of this helps the political parties to move into a low-carbon position.

If you can find a way of being happier but with a smaller footprint, you are a leader.

The cost efficiency of selected carbon-saving options

The list I give below isn't complete but I have included it to illustrate that it is essential to pick our battles. Some of the least cost-effective options on this list are receiving major UK government funding while some of the best-looking options haven't had serious attention. There could be other well-founded reasons for this, but they aren't yet obvious to me.

It can be frustrating to see public money wasted on red herrings, apparently because the analysis simply hasn't been done. Quantified carbon and cost analysis may not be the whole story, but it is an essential part of it.

All the figures below are net costs or profits over the lifetime of the measure. They are based on a financial discount rate of 10% (see Photovoltaic panels, page 133). In other words, if you are promised a saving of £1000 but have to wait a year for it, I've only called it £900. If you have to wait 2 years, I've called it £810, and so on.

- **Putting 270 mm loft insulation in homes that haven't got any**
  £70 net profit per tonne saved. £2.80 for every £1 invested.

- **Investing in offshore and onshore wind farms**
  Just above zero. Payback in 15 years (would be 8 years if we ignore discount rates). Lifetimes of the farms vary.

- **Slowing down from 70 miles per hour to 60 miles per hour on the motorway**
  Variable, but typically cost neutral even when the value of the driver's time is included. No investment costs (see page 67).

- **Pay farmers to keep their forests via the Amazon Fund or similar**
  £3 per tonne, plus biodiversity benefits (see Deforestation, page 154).

- **Funding family planning in the developing world**
  £4 per tonne according to the Optimum Population Trust. See Having a child, page 151.

- **Upgrading loft insulation to 270 mm where 50 mm currently exists**
  £5 per tonne. This figure is the total cost, which is shared between government and homeowner.

- **Government investing 24p per unit to a feed-in tariff for micro wind turbines**

£250 per tonne saved, assuming that this replaces electricity from coal, and ignoring the embodied energy in the panels themselves (see Wind turbine, page 146).

■ **Government investing 36.5p per unit to a feed-in tariff for micro-photovoltaic panels**
£360 per tonne saved, assuming that this replaces electricity from coal, and ignoring the embodied energy in the panels themselves (see Solar panel, page 133).

■ **Building to code for sustainable homes level 6 (carbon neutral) instead of to current building regulations**
Almost certainly very expensive (see A house, page 149).

Where the numbers come from

I hope I have already made the point clearly enough that carbon footprinting is a long way from being an exact process, whatever anyone ever tells you or whatever numbers you might see written on the side of products in some shops. All my numbers are best estimates and nothing more, even though I have reached them as carefully as I can.

I have tried to be as transparent as I can within the practical constraints of the book and my resources. Occasionally the sources are confidential to clients of mine, but more often it is simply too laborious to document every last detail. Nevertheless, there is a reasonable degree of transparency most of the time, and here is a summary of my approach.

I have used a variety of different methods and sources. I have drawn on a range of publicly available data sets and models, from life cycle studies and reports, and from studies I have carried out myself for businesses across different industries. I have used models that we are developing all the time in my company, Small World Consulting.

Often I've arrived at numbers from a couple of different routes to check that the results agree with each other. I've tried to put notes and references in the text wherever possible. Occasionally, frankly, it has been more a case of putting my finger in the air and guessing, but when that has been the case I've tried to make it clear.

Here are some of the main sources I have used.

Publicly available data sets drawn from process life cycle analyses

Process based life cycle analysis is the most common approach to carbon footprinting. It is often referred to as 'bottom-up' because you start off down on your hands and knees, identifying one by one all the processes that have had to happen in order for, say, a product to be created. Then you add up the emissions from each process and that's the footprint of the product. Simple! Except that it isn't. Not at all. It's back-breaking work and since the number of processes you really need to count up is always infinite, the job is never quite complete, so you end up with an underestimate. In fact the leaks are often shocking, 50% or more. To make matters worse, these problems are popularly overlooked, even in the development of government-backed and funded guidelines, such as the PAS 2050 standard (which was published despite a government-commissioned study that concluded that the draft methodology wasn't fit for some of its key intended purposes[2]).

For all the problems, and despite being hard work, process life cycle analysis is still an essential source of detailed information that can't be gathered any other way. Here are some of the key sources of this type that I've used, each of which is referenced in the main text:

- Defra publishes emissions factors for a range of fuels, electricity sources, transport modes, utilities and waste. These are mostly UK specific and don't take account of full supply chains. I use them where I can but supplement with additions for the missing supply chains.

- The University of Bath produces the Inventory of Carbon and Energy, a publicly available data set of carbon emissions factors for hundreds of materials, mainly relating to the construction industry, up to the factory gate.

- The Association of European Plastics Manufacturers (APME) publishes data sets of emissions factors for a wide range of plastics based, not surprisingly, on European manufacture.

- The UK's Market Transformation Programme has a wide range of data on the carbon intensity of common appliances.

- I have drawn on a further wide range of life cycle analysis studies from all kinds of sources. This is tricky because they all draw their boundaries in slightly different ways and use slightly different assumptions. At its best this has involved me in picking through high-quality academic studies. At its worst it has degenerated into 'Google footprinting': scrounging around the web, digging for numbers. When I've sunk to these depths, I've let you know.

Environmental input–output analysis

This is a neat alternative and complement to process life cycle analysis. It's not as popular, perhaps because it's a bit harder to get your head around, but it's at least as robust as anything else in the murky world of carbon footprinting. It is sometimes called a 'top-down' approach because it starts by looking at the whole economy from a height. It uses macroeconomic modelling to understand the way in which the activities of one industry trigger activities and emissions in every other industry. Input–output's key 'trick' is a piece of funky maths (for which a man called Wassily Leontief got a Nobel Prize) that succeeds in the capturing the endless ripple effects in a way that is 100 per cent complete. It has the further advantage that if you know how much you spend on something you can get an instant crude estimate of its carbon footprint. It's like a magic trick. And just like all the best magic it is also a bit too good to be true: the downside of input–output analysis is that the results can be ridiculously generic.

Input–output analysis is powerful tool both because it doesn't 'leak' and because once the model has been built it is often easy to use. The basic technique is well established. The specific model I've used is one we developed at Small World Consulting with Lancaster University. It draws mainly on data from the UK's Office of National Statistics. Our model is based on a 2007 picture of the UK economy; it deals with all the greenhouse gases and employs an emissions weighting factor for high-altitude emissions. A key weakness, which I refer to from time to time and sometimes adjust for, is that it treats imports as though they had the same carbon intensity as domestic production, whereas in reality they are usually more carbon intensive.

Most of the time I have used a combination of process-based and input–output approaches to get my numbers. At their best, process-based methods can be more precise, but input–output analysis is often able to get at places that process life cycle analysis is unable to reach. Putting the two methods together is sometimes called a hybrid approach, and the result is a bit like looking through both a microscope and a telescope at the same time. They each show you different things and between them, if the lenses are clean, you might end up with a passable understanding of whatever it is you are looking at.

Booths supermarkets' greenhouse gas footprint model

Over the last three years my company has been mapping out the carbon footprint of the Booths group of supermarkets and its supply chains. The model we now have draws on a great many life cycle studies of foods up to the farm gate, often using those funded by Defra. Reports and agricultural models from Cranfield University deserve a mention because I've used them extensively even though they are not uncontentious. Also well worth a mention are five reports produced by the Food Climate Research Network. The Booths model includes transport, processing, packaging refrigeration and the supermarket chain's other operations. All of these components are attributed to products, broken down into 75 categories. The model goes into a lot of detail, but that doesn't make it accurate. Human understanding of emissions from agriculture is still poor. The model is simply the best picture we have managed to achieve so far. Its purpose is purely practical and we think it is now good enough to work from, enabling actions to be reasonably well targeted on the hotspots. It is, I think, the most comprehensive model of the climate impacts of supermarket food in the public domain.

Direct greenhouse gas (GHG) emissions per GDP and per person for 60 countries

Note that these figures do not take account of embodied emissions of imported or exported products, or of international transport. They are simply estimates of the emissions that actually arise from each country.

Country	Population (millions)	GDP (billions of $)	GHG (million tonnes CO_2e)	GHG per person	GHG/ GDP$	GHG/ GDP as a percentage of UK figure	Electricity emissions intensity (kg CO_2e per kilowatt-hour)
Argentina	38,730	469	316	0.0082	0.6738	174	0.2750
Australia	221,210	561	529	0.0024	0.9430	244	0.8680
Austria	8180	242	91	0.0111	0.3760	97	0.2240
Belarus	9820	63	74	0.0075	1.1746	304	0.2940
Belgium	10,420	298	140	0.0134	0.4698	121	0.2740
Brazil	183,910	1385	983	0.0053	0.7097	183	0.0780
Bulgaria	7760	58	68	0.0088	1.1724	303	0.4720
Canada	31,950	919	758	0.0237	0.8248	213	0.2240
China	1,303,040	7219	6467	0.0050	0.8958	232	0.7710
Columbia	44,920	300	160	0.0036	0.5333	138	0.1530
Croatia	4440	50	29	0.0065	0.5800	150	0.3790
Cyprus	830	17	8	0.0096	0.4706	122	0.8430

Country	Popula-tion (millions)	GDP (billions of $)	GHG (million tonnes CO_2e)	GHG per person	GHG/ GDP$	GHG/ GDP as a percent-age of UK figure	Electricity emissions intensity (kg CO_2e per kilowatt-hour)
Czech Republic	10,210	182	147	0.0144	0.8077	209	0.5020
Denmark	5400	159	68	0.0126	0.4277	111	0.3560
Estonia	1350	18	21	0.0156	1.1667	302	0.7220
Finland	5230	144	81	0.0155	0.5625	145	0.2970
France	62,180	1626	563	0.0091	0.3462	90	0.0820
Germany	82,500	2146	1015	0.0123	0.4730	122	0.4999
Greece	11,060	226	138	0.0125	0.6106	158	0.7770
Hungary	10,110	156	83	0.0082	0.5321	138	0.4210
Iceland	290	9	3	0.0103	0.3333	86	0.0010
India	1,079,720	3115	1744	0.0016	0.5599	145	0.9120
Indonesia	217,590	722	470	0.0022	0.6510	168	0.7720
Iran	67,010	463	583	0.0087	1.2592	326	0.5230
Ireland	4060	145	68	0.0167	0.4690	121	0.5920
Italy	58,130	1491	583	0.0100	0.3910	101	0.5240
Japan	127,690	3435	1355	0.0106	0.3945	102	0.4410
Kazakhstan	14,990	103	211	0.0141	2.0485	530	1.1160
Korea (South)	48,080	906	527	0.0110	0.5817	150	0.4370
Latvia	2310	25	11	0.0048	0.4400	114	0.1810
Liechtenstein	33	2		–	–	0	
Lithuania	3440	41	20	0.0058	0.4878	126	0.1210
Luxembourg	450	29	11	0.0244	0.3793	98	0.3250
Malaysia	24,890	235	154	0.0062	0.6553	169	0.4920
Malta	400	7	3	0.0075	0.4286	111	0.8140
Mexico	104,000	935	520	0.0050	0.5561	144	0.5760
Netherlands	16,490	476	218	0.0132	0.4580	118	0.440
New Zealand	4080	87	75	0.0184	0.8621	223	0.1780
Nigeria	128,710	137	232	0.0018	1.6934	438	0.4460
Norway	4590	162	55	0.0120	0.3395	88	0.0090
Pakistan	152,060	311	230	0.0015	0.7395	191	0.370

Country	Population (millions)	GDP (billions of $)	GHG (million tonnes CO_2e)	GHG per person	GHG/ GDP$	GHG/ GDP as a percentage of UK figure	Electricity emissions intensity (kg CO_2e per kilowatt-hour)
Papua New Guinea	4809	13	7	0.0015	0.5385	139	
Poland	38,180	455	388	0.0102	0.8527	220	0.6620
Portugal	10,520	189	85	0.0081	0.4497	116	0.4140
Romania	21,690	169	155	0.0071	0.9172	237	0.4510
Russian Federation	143,850	1309	1938	0.0135	1.4805	383	0.3290
Saudi Arabia	23,950	304	371	0.0155	1.2204	316	0.7490
Slovakia	5380	72	51	0.0095	0.7083	183	0.2550
Slovenia	2000	38	20	0.0100	0.5263	136	0.3630
South Africa	45,150	468	505	0.0112	1.0791	279	0.8530
Spain	42,690	983	428	0.0100	0.4354	113	0.3810
Sweden	8990	244	70	0.0078	0.2869	74	0.0590
Switzerland	480	224	53	0.1104	0.2366	61	0.030
Thailand	63,690	474	320	0.0050	0.6751	175	0.5280
Turkey	71,790	511	304	0.0042	0.5949	154	0.4960
UK	59,840	1696	656	0.0110	0.3868	100	0.4730
Ukraine	47,450	279	414	0.0087	1.4839	384	0.3410
US	293,950	10,708	7065	0.0240	0.6598	171	0.5750
Venezuela	26,130	145	237	0.0091	1.6345	423	0.2450
EU (except UK) average					0.47053	122	
Non-EU average					0.72497	187	

Source: derived from factsheets within Höhne, N., Phylipsen, D. & Moltmann, S. (2007) *Factors Underpinning Future Action: 2007 Update*. A report by Ecofys for the Department for Environment, Food and Rural Affairs. Ecofys GmbH, Cologne. Available at <http://www.fiacc.net/data/fufa2.pdf>.

The carbon footprint of UK products and services from different industries per £ of value

The following figures come from an input–output model that used 2007 data from the Office of National Statistics and adjusted for 2009 prices. An emissions weighting factor of 1.9 for aviation has been used. The specific model used was developed and managed by Small World Consulting Ltd, in collaboration with Lancaster University.

Industry	Kilograms CO_2e per £ of output at retail prices	Kilograms CO_2e per £ of output at industry prices
Agricultural produce	2.60	2.95
Forestry produce	0.50	0.56
Fish	0.85	1.00
Coal (not including burning it)	3.56	3.99
Oil and gas (not including burning it)	0.78	0.78
Metal ores	0.99	1.04
Stone, clay and minerals	1.06	1.17
Processed meat	1.03	1.42
Processed fish and fruit	0.75	0.99
Processed oils and fats	0.62	0.75
Dairy products	1.22	1.91
Grain milling and starch	1.05	1.33
Animal feed	0.91	1.12
Bread, biscuits, etc.	0.63	0.71
Sugar	1.04	1.37
Confectionery	0.35	0.51
Other food products	0.67	0.85
Alcoholic beverages	0.26	0.35
Soft drinks and mineral waters	0.51	0.66
Tobacco products	0.13	0.15
Textile fibres	0.62	0.91
Textile weaving	0.70	0.87
Textile finishing	0.88	0.92

Industry	Kilograms CO_2e per £ of output at retail prices	Kilograms CO_2e per £ of output at industry prices
Made-up textiles	0.26	0.45
Carpets and rugs	0.21	0.39
Other textiles	0.58	0.85
Knitted goods	0.71	0.81
Clothing	0.23	0.42
Leather goods	0.51	0.71
Footwear	0.23	0.39
Wood and wood products	0.76	0.78
Pulp, paper and paperboard	1.29	1.43
Paper and paperboard products	0.71	1.01
Printing and publishing	0.35	0.41
Petroleum products, coke and nuclear fuel (not including burning)	0.64	0.65
Industrial gases and dyes	1.42	1.74
Inorganic chemicals	1.20	1.82
Organic chemicals	1.49	1.65
Fertilisers	3.05	4.26
Plastics and synthetic resins, etc.	1.24	1.35
Pesticides	1.03	1.36
Paints, varnishes, printing ink, etc.	0.56	0.74
Pharmaceuticals	0.27	0.33
Soap and toilet preparations	0.26	0.43
Other chemical products	0.63	0.72
Man-made fibres	2.07	2.22
Rubber products	0.89	1.07
Plastic products	0.89	0.95
Glass and glass products	0.94	1.17
Ceramic goods	0.46	0.65
Structural clay products	0.94	1.36
Cement, lime and plaster	4.04	5.88
Articles of concrete, stone, etc.	1.31	1.39
Iron and steel	2.63	3.24
Non-ferrous metals	1.80	2.12
Metal castings	2.86	1.27
Structural metal products	1.11	1.18

Industry	Kilograms CO_2e per £ of output at retail prices	Kilograms CO_2e per £ of output at industry prices
Metal boilers and radiators	0.71	0.79
Metal forging, pressing, etc.	1.49	0.94
Cutlery, tools, etc.	0.46	0.62
Other metal products	0.94	1.14
Mechanical power equipment	0.71	0.76
General purpose machinery	0.75	0.77
Agricultural machinery	0.69	0.79
Machine tools	0.59	0.66
Special purpose machinery	0.85	0.95
Weapons and ammunition	0.59	0.60
Domestic appliances	0.44	0.64
Office machinery and computers	0.36	0.42
Electric motors, generators, etc.	0.64	0.70
Insulated wire and cable	1.07	1.39
Other electrical equipment	0.47	0.58
Electronic components	0.48	0.53
Transmitters for TV, radio and phone	0.44	0.47
Receivers for TV and radio	0.23	0.36
Medical and precision instruments	0.33	0.39
Motor vehicles	0.72	0.82
Shipbuilding and repair	0.70	0.76
Other transport equipment	0.39	0.54
Aircraft and spacecraft	0.56	0.56
Furniture	0.52	0.66
Jewellery and related products	0.40	0.64
Sports goods and toys	0.18	0.41
Miscellaneous manufacturing and recycling	0.53	0.76
Electricity production and distribution	6.12	6.12
Gas distribution	1.65	1.65
Water supply	0.67	0.67
Construction	0.38	0.38
Motor vehicle distribution and repair, fuel	–	0.32
Wholesale distribution	–	4.79

Industry	Kilograms CO_2e per £ of output at retail prices	Kilograms CO_2e per £ of output at industry prices
Retail distribution	–	2.89
Hotels, catering, pubs, etc.	0.37	0.37
Railway transport	0.89	0.89
Other land transport	0.85	0.85
Water transport	2.07	2.07
Air transport	4.59	4.59
Ancillary transport services	0.28	0.28
Postal and courier services	0.38	0.38
Telecommunications	0.25	0.25
Banking & finance	0.16	0.16
Insurance and pension funds	0.22	0.22
Auxiliary financial services	0.15	0.15
Owning and dealing in real estate	0.10	0.10
Letting of dwellings	0.07	0.07
Estate agent activities	0.09	0.09
Renting of machinery, etc.	0.25	0.25
Computer services	0.11	0.11
Research and development	0.20	0.20
Legal activities	0.10	0.10
Accountancy services	0.12	0.12
Market research, management consultancy	0.14	0.14
Architectural activities and technical consultancy	0.15	0.15
Advertising	0.14	0.14
Other business services	0.15	0.15
Public administration and defence	0.33	0.33
Education	0.17	0.17
Health and veterinary services	0.23	0.23
Social work activities	0.27	0.27
Sewage and sanitary services	1.67	1.67
Membership organisations	0.14	0.14
Recreational services	0.21	0.21
Other service activities	0.24	0.24

Notes and references

Introduction

1 The phrase 'save the planet' is just shorthand for 'save the people on the planet'. The Earth will be fine until changes in the Sun's radiation evaporate its atmosphere in a billion or so years time. By this time, as Lord Martin Rees, president of the Royal Society, speculated, the creatures that inhabit the Earth will be as different from people as we are from bacteria.

A quick guide to carbon and carbon footprints

1 *Carbon Footprinting: An Introduction to Organisations*, published by the UK's Carbon Trust (2007) defines on page 1 a carbon footprint in a similar way to me but goes on to describe 'basic carbon footprints' on page 4. These are toe-prints rather than rough estimates of footprints.

2 All the gases covered by the Kyoto Protocol are included. For better or worse, I have adopted a common convention of considering the impact over a 100-year period. This makes a difference because some gases last longer than others. CO_2 stays in the atmosphere for a very long time, whereas methane and refrigerant gases fade much faster. If we were measuring the impact over 20 years, methane would be about three times more prominent.

3 There is much scientific uncertainty around the impact of high-altitude emissions. A figure of 1.9 can be inferred from the IPCC 4 assessment report. This is also the figure suggested in the 2009 Guidelines to Defra/DECC's Greenhouse Gas Conversion Factors for Company Reporting (Annex 6, footnote 10).

4 How many tonnes for a life? – a back-of-the-envelope calculation. I'm not presenting this as even the beginnings of a rigorous argument, but just as a line of thought that might serve to make that critical connection between carbon and life itself.

Let's see whether it's possible to estimate of the number of tonnes of carbon released into the sky before it would be reasonable to think that someone somewhere is going to have to die as a consequence.

It will be simplistic but I hope you will stick with it. I'm going to take a high-emissions scenario and a low-emissions scenario for the world and make assumptions about how many people will die as a result of climate change from each. Then I'll take the difference in the carbon emissions, divide it by the difference in the number of deaths around the world and use that as the guide to the number of tonnes per death.

For my low-emissions scenario, I'm going to assume that we immediately cut global emissions by 40 per cent to 30 gigatonnes CO_2e – that's 30 billion tonnes – and hold them like that for 40 years. (Clearly that is impossible. I'm going to use this scenario because the sums are simple. A more realistic scenario with a similar climate outcome might be a rapid reduction starting today and resulting in global emissions falling to about 10 gigatonnes by 2050. This would still be an extremely radical response and one that the world hardly seems on the brink of adopting.)

Let's assume that under the low-emissions scenario relatively few people – a small fraction of a billion – die as a result of climate change. That looks likely, although there is still a risk that the outcome would be worse. One estimate is that the death rate related to climate change is already 300,000 per year. We know that the climate is going to continue to get warmer for years even if we cut our emissions to zero right now, because of the greenhouse gases that are already in the atmosphere, so that 300,000 figure is unlikely to reduce. Still, it seems entirely plausible to keep things down to such a small fraction of a billion that for the purpose of this estimate we can call it zero.

For my high-emissions scenario I'm going to assume that until 2050 we average the current 50 gigatonnes CO_2e per year. Many scientists think our species will be in a **lot of trouble** if that happens. It's not unrealistic to think that we might average the current 50 gigatonnes when you think that right now the trend in global emissions is not just rising fast but still accelerating (tempered only by the short-term blip of the global recession). One view is that in that kind of scenario, by 2100 the world will only be able to support 1 billion people instead of the 9 billion of us that are forecast without climate change.

(To visualise the time lags involved in reversing climate change, imagine someone in a car, accelerating like crazy and already way above the speed limit when they notice that they are in an 'average speed check area'. To avoid the fine, the driver has to first switch his foot from accelerator to brake, then wait for the car to slow down and *then* drive slowly for quite a while until their *average* speed slips below the limit. Each step in the process takes time. It's the same for the world, only far worse. The rate of emissions corresponds to the power going into the engine, but it takes years for the global community even to start wondering whether to begin the process of adjusting the position of its feet on the pedals. Once that has been decided and the foot has been shifted, the long and difficult process of braking can begin. Even when, many years later, the stuff coming out of our chimneys, fields and forest fires around the world has dropped to a low enough level, the temperature will carry on rising until the total amount of carbon in the reservoir of the sky has dropped by enough for the temperature to begin to fall. Even then any remaining ice will carry on melting until the temperature has actually dropped. Whatever we do now, falling temperatures are decades away, and no one reading this book is likely to see increasing ice. And right now we are still a long way from having our foot on the right pedal. I've over-simplified the science, but the basic concept is right.)

So that is a difference of 8 billion people. (The Optimum Population Trust estimates a mere 5 billion difference, but their figures are in the same ball park as mine: 'Earth heading for 5 billion overpopulation?' (Optimum Population Trust, March 2009), <**http://www.optimumpopulation.org/ releases/opt.release16Mar09.htm**>.) It is not fair to say that all those 8 billion will have to die as a result of climate change, because some will simply never be born, so let's say that in this scenario, 4 billion people will die. It's chilling, and you may not buy into the argument completely. None of us knows exactly what would happen, and we can't actually run both scenarios to find out. But I'm going to run with these numbers for the sake of the thought experiment.

The difference between the scenarios is 600 billion tonnes of CO_2e and 4 billion deaths. That works out at one death per 150 tonnes CO_2e.

5   £12 per tonne CO_2e is the maximum price that a company might have to pay under the Carbon Reduction Commitment. The European Trading Scheme puts it at about half of that.

Under 10 grams

1   Drawn from a Swedish life cycle assessment study in 2004 by Mireille Faist Emmenegger, Rolf Frischknecht, Markus Stutz, Michael Guggisberg, Res Witschi and Tim Otto: 'Life Cycle Assessment of the mobile communication system UMTS: towards eco-efficient systems', <**http://www.esu-services.ch/ download/faist-2005-umts.pdf**>.

2   Gartner press release, 2008: 'Gartner says mobile messages to surpass 2 trillion messages in major markets in 2008', <**http://www.gartner.com/it/ page.jsp?id=565124**>. With 1.9 trillion messages in 2007, they predicted 2.3 trillion in 2009. I have extrapolated a bit further.

3   Based on 150 litres per person per day, which is about average in the UK. Defra domestic water consumption summary, <**http://www.defra.gov.uk/ sustainable/government/progress/regional/summaries/16.htm**>.

4   'The energy needed to treat and pump mains water to our homes, and to collect and treat waste water from the sewage network, is responsible for

nearly 1 per cent of the UK's annual greenhouse gas emissions.' UK's Energy Saving Trust, 'Water and carbon – the facts', <**http://www.energysaving-trust.org.uk/Water/Water-and-carbon-the-facts**>. EST's figure for total UK emissions excludes the overseas component of the footprint. Small World Consulting's input–output model puts the greenhouse gas emissions from household water supply at 0.3 per cent and emissions from sewage and sanitary services at nearly 2 per cent, but that includes other things.

5 Based on figures for the carbon intensity of UK water supply and treatment: Defra (2009), 'Guidelines to Defra's GHG conversion factors for company reporting', <**http://www.defra.gov.uk/environment/business/reporting/pdf/20090928-guidelines-ghg-conversion-factors.pdf**> [accessed February 2010].

6 Word for word, the Google story, from their blog, is: 'Queries vary in degree of difficulty, but for the average query, the servers it touches each work on it for just a few thousandths of a second. Together with other work performed before your search even starts (such as building the search index) this amounts to 0.0003 kW h [kilowatt-hours] of energy per search, or 1 kJ [kilojoule]. For comparison, the average adult needs about 8000 kJ a day of energy from food, so a Google search uses just about the same amount of energy that your body burns in ten seconds. In terms of greenhouse gases, one Google search is equivalent to about 0.2 grams of CO_2.' <**http://google-blog.blogspot.com/2009/01/powering-google-search.html**>

7 This estimate is based on a breakdown of power usage at Lancaster University.

8 For example, Dr Alex Wissner-Gross, a physicist from Harvard University, wrote in *The Sunday Times* on 9 January 2009: 'How you can help reduce the footprint of the Web', <**http://www.timesonline.co.uk/tol/news/environment/article5488934.ece**>.

9 The sums: 5 m² doorway, fully open for 15 seconds, wind speed through the door of 1 metre per second, temperature difference of 15°C, heat capacity of air 1.2 kilojoules per cubic metre, heat supplied by gas at 0.22 kg CO_2e per kilowatt-hour.

10 BREEAM: Building Research Establishment Environmental Assessment Method. I understand that the BRE has since improved its energy efficiency criteria somewhat. The sums here are based on a temperature difference of 15°C (typical for winter) and a wind speed of just 2.5 miles per hour flushing warm air out of the building.

11 I am assuming that this low-grade paper comes in at just 1 kg CO_2e per kilo.

12 Association of Plastic Manufacturers. Eco-profiles showing emissions from production of a wide variety of plastics are available from <**http://lca.plasticseurope.org/index.htm**> [accessed 20 April 2008]. Based on 3 g weight per bag.

13 <**http://www.reusablebags.com/facts.php**>. Vincent Cobb's website

(<http://www.reusablebags.com>) contains interesting data on the numbers of bags used around the world, as well as their impacts.

10 grams *to* 100 grams

1 Annex 9 in Defra (2009), 'Guidelines to Defra's GHG conversion factors for company reporting' (<**http://www.defra.gov.uk/environment/business/ reporting/pdf/20090928-guidelines-ghg-conversion-factors.pdf**>), gives a figure of 550 kg CO_2e per tonne of paper and card in landfill.

2 <**www.caloriesperhour.com**>.

3 In our input–output model of the greenhouse gas footprint of UK industries, sports goods typically have a carbon intensity of around 250 g per pound's worth of goods at retail prices. If we make the very broad assumption that cycling goods are typical of this, and if we say that Her Majesty's Revenue and Customs (HMRC) is being roughly fair to reimburse you at 20p per mile for business travel on a bike, then we would need to add about 50 g CO_2e per mile to take account of the wear and tear on your bike, your waterproofs, lights, helmet and so on. Actually, as someone one who is frequently cycling between offices and train stations trying to keep jacket, tie and laptop dry, I suspect that HMRC has underestimated it and should be paying out the full 40p per mile that they allow for car users. (This would also provide a beneficial incentive.)

4 Saunders, C., Barber, A. & Taylor, G. (2006) *Food Miles – Comparative Energy/Emissions Performance of New Zealand's Agriculture Industry.* Research Report no. 285. Lincoln, New Zealand: Lincoln University.

5 Blanke & Burdick (2005), referenced in Defra (2006) *Environmental Impacts of Food Production and Consumption*, p. 47, <**http://randd.defra.gov.uk/ Document.aspx?Document=EV02007_4601_FRP.pdf**>.

6 For the footprint up to the farm gate I've used a number from Wallen, A., Brant, N. & Wennersten, R. (2004) Does the Swedish consumer's choice of food influence greenhouse gas emissions? *Envir Sci Policy* 7, 525–535. For the rest of the footprint I've used my work at Booths.

7 Oxfam Cool Planet website has a simple (and child-friendly) account of how Fair Trade bananas are grown in the Windlass Islands along with recipes: <**http://www.oxfam.org.uk/coolplanet/kidsweb/banana/index. htm**>.

8 There is more on this at the very accessible Banana Link website 'Working towards a fair and sustainable banana trade', <**http://www.bananalink.org. uk**>. For a critical and pessimistic look at the future of bananas in our lives see also Dan Koeppel (2008) 'Yes, we will have no bananas', *New York Times* (18 June), <**http://www.nytimes.com/2008/06/18/opinion/18koeppel. html?_r=1**>. Waitrose has commissioned a life cycle analysis of one of their banana supply chains through the University of Bangor, due for publication

later in 2010, I gather. I understand that this has taken account of deforestation issues and will therefore make an interesting read when it becomes available.

9 My numbers for the footprint up to the farm gate come from two sources: a one of oranges grown in Spain, the other of produce for the Swedish market: Sanjuan, N., Ubeda, L., Clemente, G. & Mulet, A (2005) LCA of integrated orange production in the Comunidad Valenciana (Spain). *Int J Agric Resources Governance Ecol* 4 (2), 163–177; and Wallen, A., Brant, N. & Wennersten R. (2004) Does the Swedish consumer's choice of food influence greenhouse gas emissions? *Envir Sci Policy* 7, 525–535. For the rest of the footprint I've used our work at Booths.

10 Ofcom news release, 17 December 2009: 'The UK witnessed the highest average increase in TV watching during 2008, up by 3.2 per cent to 3.8 hours a day. This was higher than the average (3.5 hours per day) across the European countries surveyed, but still slightly less than viewers in Italy, Poland and Spain. US viewers consumed the most television in 2008, watching on average 4.6 hours a day, up 1.8 per cent from 2007, whilst viewers in Sweden continued to watch the least at 2.7 hours a day, although this was a 1.9 per cent increase across the year.' <**http://www.ofcom.org.uk/media/news/2009/12/nr_20091217**>.

11 All the data on TV power consumption come from Ireland's Electricity Supply Board (2009), <**http://www.esb.ie/main/sustainability/energy-services.jsp**>. I have allowed 10 g per hour for the satellite receiver.

100 grams *to* 1 kilo

1 The London bus occupancy factor is from the Transport for London 2008 environment report available at <**http://www.tfl.gov.uk/assets/downloads/corporate/environment-report-2008.pdf**> and <**http://www.tfl.gov.uk/assets/downloads/corporate/environment-report-2008-data-tables.pdf**>.

2 Aumônier, S., Collins, M. & Garrett, P. (2008) *An Updated Lifecycle Assessment Study for Disposable and Reusable Nappies*, Science Report no. SC010018/SR2. UK Environment Agency. <**http://randd.defra.gov.uk/Document.aspx?Document=WR0705_7589_FRP.pdf**> [accessed 1 March 2010].

3 Data from Garnett, T. (2006). *Fruit and Vegetables & UK Greenhouse Gas Emissions: Exploring the Relationship.* Food Climate Research Network, Surrey; and Foster, C., Green, K., Bleda, M., Dewick, P., Evans, B., Flynn, A. & Mylan, J. (2006) *Environmental Impacts of Food Production and Consumption: A Report to the Department for Environment, Food and Rural Affairs.* Manchester Business School. Defra, London.

4 Direct emissions from fuel and electricity generation and supply come from Defra (2009) *Guidelines to Defra/DECC's GHG Conversion factors for*

Company Reporting. Produced by AEA for the Department of Energy and Climate Change (DECC) and the Department for Environment, Food and Rural Affairs (Defra). Supply chains and infrastructure are estimates made from the input–output model.

5 David J. C. MacKay lays out the maths nicely in *Sustainable Energy Without the Hot Air* (2009), published by UIT Cambridge Ltd; available as a free download on <**www.withouthotair.com**>.

6 Kemp, R. (2007) *Traction Energy Metrics.* Rail Safety & Standards Board, London. Available from <**http://www.rssb.co.uk/pdf/reports/research/ T618_traction-energy-metrics_final.pdf**>.

7 The carbon intensity of PET, from which bottles are typically made, is 3.3 kg CO_2e per kilo, based on figures from the Association of Plastic Manufacturers, *Eco-profiles of the European Plastics Industry – Main Flow Chart*, available from <**http://lca.plasticseurope.org/index.htm**> [accessed 20 April 2008]. The bottles I have weighed average around 50 g per litre of capacity. The transport footprint is based on a UK typical rigid lorry, and conversion factors also from DEFRA – and is therefore light by up to 50 per cent because fuel supply chains and the embodied footprint of the vehicle are not taken into account. I took the energy of bottle manufacture to be around one-quarter of the energy required to make the PET pellets, on the basis of Stefano Botto, 'Tap water vs. bottled water in a footprint integrated approach' (July 2009), <**http://precedings.nature.com/documents/3407/ version/1**>.

8 The consumption figures came from Andrea Thompson, 'The energy footprint of bottled water', Live Science, 19 March 2009 <**http://www.live-science.com/environment/090318-bottled-water-energy.html**> [accessed October 2009]. According to this, most is consumed in the US and Europe, with the US accounting for 33 billion litres (110 litres per person).

9 Confederation of Paper Industries (2006) *UK Paper Making Industries Statistical Facts Sheet;* Utrecht Centre for Energy Research (2001) ICARUS-4: *Sector Study for the Paper and Board Industry and the Graphical Industry*, available from <**http://copernicus.geog.uu.nl/uce-uu/downloads/Icarus/ Paper.pdf**> [accessed 3 April 2008].

10 Defra (2009) 'Guidelines to Defra's GHG conversion factors for company reporting' (<**http://www.defra.gov.uk/environment/business/reporting/ pdf/20090928-guidelines-ghg-conversion-factors.pdf**>), Annex 9.

11 All the numbers on waste impacts come from Defra Annex 9 (see note 9 above), released in September 2009, and are based on the UK. Conversion factors for virgin and recycled paper came from Confederation of Paper Industries (2006) (see note 9 above). Environmental Defence Fund (1995) 'Energy, air emissions, solid waste outputs, waterborne wastes and water use associated with component activities of three methods for managing newsprint' provided a sense and some figures for transport and printing impacts.

12 Based on UK electricity at 0.6 kg CO_2e and gas at 0.225 kg CO_2e per kilowatt-hour. Both figures are based on those supplied by Defra (2009), but are adjusted to take account of power station supply chains and distribution. The cost of electricity is taken as 10p per kilowatt-hour.

13 Figures from Defra's Boiler Efficiency Database, 2010 <http://www.sedbuk. com/> are shown in the table below.

Boiler type	Seasonal efficiency (per cent)	Typical annual fuel cost (£)				
		Flat	Bungalow	Terraced	Semi-det'd	Detached
Old (heavy weight)	55	£267	£341	£354	£397	£550
Old (light weight)	65	£231	£293	£304	£340	£470
New (non-condensing)	78	£197	£249	£258	£289	£396
New (condensing)	88	£178	£224	£232	£259	£355

14 Based on Defra (2008) 'Guidelines to Defra's GHG conversion factors for company reporting'. Available from <http://www.defra.gov.uk/environ-ment/business/reporting/pdf/ghg-cf-guidelines-annexes2008.pdf>, supplemented with estimates derived from an environmental input–output model of emissions from supply chains that are not included in the Defra figures. These additions account for around 10 per cent of the total footprint of UK electricity.

15 Public opinion research by Landor Associates (2007) *ImagePower Green Brands Survey*, available from <http://www.landor.com/?do=news. pressrelease&r=&storyid=508>, found that whereas 'the adoption of green values is the fastest consumer trend in recent years, faster than the uptake of the internet or mobile phone', the public still lack a sophisticated under-standing of what it actually means to be green. The colour of the logo was frequently taken as one of the key indicators of green credentials.

16 This figure is drawn from my work with Booths supermarkets.

17 This is an average figure. Bargain flights can be a lot worse. If you get your London to Madrid return for £10, that's around 50 kg CO_2e per £. On the other hand, if you pay over the odds it's less carbon intensive.

18 I've used a figure of 0.68 kg CO_2e per mile for an average petrol car (including fuel supply chains and manufacture of the car). Exhaust pipe emissions from Defra (2009) *Guidelines to Defra/DECC's GHG Conversion Factors for Company Reporting. Produced by AEA for DECC and Defra.* Available from <http://www.defra.gov.uk/environment/business/report-ing/pdf/20090928-guidelines-ghg-conversion-factors.pdf>

19 Tim Jackson (2009) *Prosperity without Growth: Economics for a Finite Planet.* Earthscan, London.

20 To keep things simple I'm not going to look at the 12 per cent of general waste that gets incinerated instead of being sent to landfill. It doesn't change the overall picture much.

21 All the numbers of waste impacts come from Defra Annex 9 (see note 10 above), and are based on the UK. Data on UK consumer waste come from the Office of National Statistics web site, <**http://www.statistics.gov.uk/cci/ nugget.asp?id=1769**> [accessed October 2009]. The breakdown of consumer waste by type comes from the BBC (July 2007), 'Household waste: in statistics', <**http://news.bbc.co.uk/1/hi/uk/6222288.stm**>.

22 UK Market Transformation Programme: *BNW16: A Comparison of Manual Washing Up with a Domestic Dishwasher,* <**www.mtprog.com/spm/download/document/id/598**> [accessed October 2009].

23 I based this on a machine costing £350, lasting 10 years and used 120 times per year, and a figure of 0.44 kg CO_2e per pound sterling expenditure on domestic appliances at 2008 retail prices from the input–output model.

24 The Market Transformation Programme study (see note 22 above) points to a possible carbon saving of about one-fortieth from a decrease in water use. A whole bottle of washing-up liquid probably has a footprint of about 1 kg CO_2e.

25 <http://www.toiletpaperworld.com/>

26 Worldwatch Institute (2007) *The Reality Behind Toilet Paper Consumption,* <http://www.worldwatch.org/node/5162> [accessed October 2009].

27 Tesco press release, 1 May 2009: 'Tesco carbon labels toilet paper', <**http:// www.tescoplc.com/plc/corporate_responsibility_09/news/press_releases/ pr2009/2009-05-01/**>.

28 The figures come from models used by Small World Consulting (<**www.swconsulting.co.uk**>). An input–output approach is used for the fuel supply chains and the depreciation of the embodied emissions in the car and its manufacture.

29 Derived from Defra (2008) *Passenger Transport Emissions Factors: Methodology Paper.* Available from <**http://www.defra.gov.uk/environment/business/reporting/pdf/passenger-transport.pdf**> [accessed 3 April 2008].

30 The (slightly simplified) physics of the energy per mile required to move a car looks like this: Energy per mile = Energy required to overcome rolling resistance + Energy required to overcome air resistance. The rolling resistance component is independent of your speed, but the energy per mile required to overcome the air resistance goes up with the square of your speed. At motorway speeds the rolling resistance fades out of the picture in comparison with the air resistance, so the total energy per mile, and hence the fuel consumption, becomes proportional to the square of the speed.

Therefore dropping your speed by one-seventh gets you an improvement in miles per gallon of more than a quarter. See MacKay (2009) (note 5 above).

31 Williams, A. (2007) 'Comparative study of cut roses for the British market produced in Kenya and the Netherlands. Précis report for World Flowers, 12 February 2007', <**http://www.fairflowers.de/fileadmin/flp.de/Redaktion/ Dokumente/Studien/Comparative_Study_of_Cut_Roses_Feb_2007.pdf**>. The numbers for Kenyan and Dutch roses were derived from this study by Cranfield University. Note that the study was commissioned by World Flowers, which imports lots of flowers from Kenya. Note also that only the précis report was made public. We adjusted the figure for Kenyan roses down slightly because Cranfield had applied an emissions weighting factor of 2.7, in contrast with the 1.9 that we have been using in this book. On the other hand, the report uses generic air freight conversion factors supplied by Defra. A final note of caution is that the generic air freight conversion factors used do not take account of inefficiencies due to the bulkiness of flowers. Despite all these reservations, I think the broad conclusion of the report is probably right, and on the back of it, Booths is sourcing its Valentine's Day roses from Kenya instead of Holland this year.

32 Emissions up to the farm gate are from Williams, A. G., Audsley, E. & Sandars, D. L. (2006) *Determining the Environmental Burdens and Resource Use in the Production of Agricultural and Horticultural Commodities*. Main Report. Defra Research Project ISO205. Cranfield University, Bedford, and Defra. Available on <**www.silsoe.cranfield.ac.uk**> and <**www.defra. gov.uk**>. Beyond the farm gate, based on study by Small World Consulting for Booths Supermarkets, Berners-Lee, M. (2009) 'A re-measure of Booths greenhouse gas emissions arising from the operation and product supply chains' (Unpublished private research).

33 International Energy Agency (2007) *Tracking Industrial Energy Efficiency and CO_2 Emissions*, <**http://www.iea.org/work/2007/tracking/confer- ence_proceedings.pdf**>. Small World Consulting's input–output model suggests that direct emissions and electricity between them account for just 91 per cent of the footprint of the UK cement industry. I have therefore added 10 per cent to the IEA figures to take account of energy supply chains and other indirect emissions in the cement industry.

1 kilo *to* 10 kilos

1 I'm using conversion factors of 1.59, 2.59 and 2.70 kg CO_2e per kilo for printing on recycled, typical UK mix and 100 per cent virgin paper. Confed- eration of Paper Industries (2006) UK Paper Making Industries Statistical Facts Sheet, <**http://www.paper.org.uk/info/reports/fact2006colour0707. pdf**> [accessed 3 April 2008].

2 It's difficult to talk about the footprint of a new product innovation, because

it all depends on how much of the R&D and tooling up you assign to the first models. They currently sell at around £150, but those prices will probably tumble. The input–output model gives us a carbon intensity of around 0.6 kg CO_2e per unit output for the computer industry. That's a footprint of 90 kg CO_2e, but a lower figure of 50 kg is probably more realistic, reflecting the impact of mass production.

3   From Wrap (2008) *The Food We Waste*. Waste & Resources Action Programme (WRAP), Banbury. Available at <**http://www.wrap.org.uk/ downloads/The_Food_We_Waste_v2__2_.ec417f6e.5635.pdf**>. The overall figure of 30 per cent waste includes bones and other bits that we don't all consider edible.

4   All the numbers here come from the Association of Plastic Manufacturers, **Eco-profiles of the European Plastics Industry**, available from <**http://lca. plasticseurope.org/index.htm**> [accessed 20 April 2008].

5   2K Manufacturing are scheduled to start bulk production in March. I am typing amid the 2009 pre-Christmas snow – another example of the strange time warp that exists between us.

6   A typical bath can hold about 120 litres (with you in it too). I've taken the cold water temperature to be 8°C and a comfortable bath temperature to be 39°C. The heat capacity of water is 4.2 kilojoules per litre. I've assumed a 90 per cent efficient boiler and used a conversion factor of 0.225 kg CO_2e per kilowatt-hour for heat produced by natural gas (this uses a figure from Defra for the direct emissions of burning gas and adds to that a figure from our input–output model to estimate the supply-chain impacts). There are 3600 kilojoules per kilowatt-hour. The footprint of the bath in kg CO_2e is $120 \times (39 - 8) \times 4.2 \times 0.225/(3600 \times 90$ per cent$)$. The footprint of the water consumption is negligible.

7   At the time of writing, I'm told that Booths is switching its bobby beans from Peru to nearer-by Egypt. It's still air freight, but a big reduction none the less.

8   Input–output modelling gives a guideline of 440 g CO_2e per pound sterling's worth of expenditure on domestic appliances. I have assumed that each is good for 1000 uses, and based my calculation on washing machines costing £300, and tumble driers £200.

9   The rating system seems a bit unfair on condensing driers since it doesn't take account of the fact that they keep the heat in the home instead of belching it into the outside world.

10 Based on 2500 calories for a man, and 2000 calories for a woman.

11 Williams, A. G., Audsley, E. & Sandars, D. L. (2006) *Determining the Environmental Burdens and Resource Use in the Production of Agricultural and Horticultural Commodities*. Main Report. Defra Research project IS0205. Cranfield University, Bedford, and Defra. Available from <**www. silsoe.cranfield.ac.uk**> and <**www.defra.gov.uk**>.

12 Best guess based on the 'average figure', which I have derived simply as a round number midway between efficient and inefficient production.

13 International Rice Research Institute, <**http://beta.irri.org**>.

14 International Rice Research Institute, <**http://beta.irri.org**>. See also 'Fertiliser', page 138.

15 Rice statistics are shown in the table below. Sources: International Rice Research Institute, <**http://beta.irri.org**>; Worldwatch Institute (2009) *State of the World 2009: Confronting Climate Change*, 26th edn. Earthscan, London. Total consumption figure taken from 2008, total fertilizer figure from 2005.

	Low estimate	High estimate
Global rice consumption (million tonnes)	432	432
Fertiliser applied (million tonnes)	161	161
Percentage of global calories provided	19.7	19.7
Annual methane from paddy fields (million tonnes CO_2e)	600	600
Emissions per kilo of fertiliser	2.7	12.3
Emissions from fertiliser use	435	1984
Transport and other agricultural emissions (million tonnes CO_2e)	43	43
Footprint per kilo of rice (kg CO_2e)	2.5	6.1
Agricultural footprint if we fed the world like this (million tonnes CO_2e)	5476	13,340

Carbon footprint of rice production

16 Data taken or derived from WWF (2007) *Making Water: Desalination: Option or Distraction for a Thirsty World?*, <**http://assets.panda.org/downloads/desalinationreportjune2007.pdf**> [accessed October 2009].

17 According to a report by the European Environment Agency, Spain and Portugal will be most affected within the EU by the coming climate change. Storms, floods and droughts are likely to become more and more frequent. Moreno, J. M. et al. (2005) *A Preliminary General Assessment of the Impacts in Spain due to the Effects of Climate Change*. Ministerio de Medio Ambiente. Available from <**http://www.mma.es/secciones/cambio_climatico/areas_tematicas/impactos_cc/pdf/evaluacion_preliminar_impactos_completo_2.pdf**>.

18 Information on Seawater Greenhouse from Wikipedia, <**http://en.wikipedia.org/wiki/Seawater_greenhouse**>. My calculations also used

figures on UK water consumption from the UK government: 'Indicators of sustainable development', <**http://www.sustainable-development.gov.uk/ sustainable/quality04/maind/04q02.htm**> [accessed October 2009], and a figure of 0.6 kg CO_2e per kilowatt-hour of electricity to drive pumps. None of my analysis includes any of the electrical energy that is required only to pump water to and from the sea. Seawater Greenhouse figures come out at 3.9 kg CO_2e per tonne, on the basis of a 300-mile journey with a 100-metre height gain, using grid electricity from a fuel mix similar to that of the UK's and not taking into account the possibility for recovering some of the energy from the brine returning downhill through turbines.

19 Kalliala, E. M. & Nousiainen, P. (1999) Life cycle assessment: environmental profile of cotton and polyester-cotton fabrics. *AUTEX Res J* **1**(1), 8–20.

20 Based on a study on cotton towels, which found that over the 2-year lifetime of the towel, the laundry was going to have three and a half times the impact of the creation of the towel. I've assumed that my jeans will last longer, but I'm nevertheless suspicious that the embodied emissions have been under-estimated in the study. I've assumed that these two will cancel each other out. Blackburn, R. & Payne, J. (2004) Life cycle analysis of cotton towels: impact of domestic laundering and recommendation for extending periods between washing. *Green Chem* **6**, G59–G61.

21 Association of Plastic Manufacturers (see note 4 above).

22 A report by the Stockholm Environment Institute estimated that it takes between 9788 and 9958 litres of water to produce 1 kg of cotton and that this represents 1.7 per cent of the UK's water footprint. Cherrett, N., Barrett, J., Clemett, A., Chadwick, M. & Chadwick, M. J. (2005) *Ecological Footprint and Water Analysis of Cotton, Hemp and Polyester.* Report prepared for and reviewed by Bioregional Development Group and World Wide Fund for Nature – Cymru. Stockholm Environment Institute, Stockholm. Available at <**http://www.organicexchange.org/Farm/Reading%20and%20Refer-ences/Cotton%20Hemp%20Polyester%20study%20SEI%20and%20Biore-gional%20and%20WWF%20Wales.pdf**>.

23 Impacts up to the farm gate are from Cranfield 2006. Impacts from the farm to the checkout are from Small World Consulting's work for Booths super-markets 2007–2009.

24 See previous note.

25 There are studies giving lower figures than this, but I have also read suspi-cions that they underestimate. Garnett, T. (2006) *Fruit and Vegetables & UK Greenhouse Gas Emissions: Exploring the Relationship.* FCRN working paper 06-01. Food Climate Research Network (FCRN). Available at <**http://www. fcrn.org.uk/fcrnPublications/publications/PDFs/Fruitnveg_paper_2006. pdf**>.

26 Williams, Audsley & Sandars (2006) (see note 10 above). Although this looks like the best information around, it is contested. I know farmers who

are highly critical of the assumptions made in the same report about organic dairly herds. My high-end figure is adjusted upwards from Cranfield's 38.6 kg CO_2e per kilo to take account of produce from a colder time of year rather than the year-round average reported in the Cranfield study.

27 The figure for trout up to the slaughterhouse comes from a database of life cycle analyses sponsored by the Danish government: Nielsen, P. H., Nielsen, A. M., Weidman, B. P., Dalgaard, R. & Halberg, N. (2003) LCA Food Database, 'Lifecycle Assessment of Basic Food' (2000–2003). Aarhus University, Denmark, <http://www.lcafood.dk/>. I have used Booths supermarkets' carbon impact model to estimate impacts from the slaughterhouse to the checkout.

28 See note 26 above. The figures for frozen filleted fish, fish products, unprocessed shellfish and processed shellfish come from Wallen, A., Brant, N. & Wennersten, R. (2004) Does the Swedish consumer's choice of food influence greenhouse gas emissions? *Envir Sci Policy* 7, 525–535. The final figure for cod comes from Foster, C. *et al.* (2006) *Environmental Impacts of Food Production and Consumption. A Report to the Department for Environment, Food and Rural Affairs.* Manchester Business School. Defra, London, p. 101. Available at <http://randd.defra.gov.uk/Document. aspx?Document=EV02007_4601_FRP.pdf>. The agreement with the Danish figures above is encouraging. Unless otherwise stated, the figures come from Nielsen *et al.* (see note 27 above).

29 Figures from the Inventory of Carbon and Energy, a publicly available data-base of embodied energy figures for several hundred materials, compiled from the best-available life cycle analyses around the world (Hammond, G. & Jones, C. (2008) *Inventory of Carbon and Energy (ICE)*, version 1.6a. University of Bath).

30 Höhne, N., Phylipsen, D. & Moltmann, S. (2007) *Factors Underpinning Future Action: 2007 Update.* A report by Ecofys for the Department for Environment, Food and Rural Affairs. Ecofys GmbH, Cologne. Available at <http://www.fiacc.net/data/fufa2.pdf>. The data here are extrapolated from Climate Fact Sheets for different nations.

10 kilos *to* 100 kilos

1 The widow of former Philippines president Ferdinand Marcos was listed by *Newsweek* as one of the 100 'greediest people of all time'. She gained some of her notoriety from her shoe collection, gathered while plenty of her fellow citizens lived in poverty.

2 A weakness of the input–output model I used for this is that it assumes that Chinese production is as carbon efficient as UK manufacture. It isn't. It's worse. In reality, a key carbon decision for footwear suppliers is where to have product made.

3   US Food Safety and Inspection Service, <**http://www.fsis.usda.gov/HELP/ FAQs_Hotline_Illness/index.asp**> [accessed October 2009].

4   This is the additional footprint arising from your decision to make the commute given that everyone else is already on the road. It is also the difference you can make by stopping commuting. It is more than your fair share of the total pollution, which would only be double rather than three times the normal emissions from driving that distance on an empty road.

5   To make it very simple, think of a queue 10 cars long, moving at one car per minute. Assuming the queue has stayed the same size, those 10 cars will between them have queued for 100 car minutes by the time they have all gone through. Add your car and you have 11 cars all queuing for 11 minutes. That's 21 minutes more queuing, even though you experience just 11 minutes. You get the same effect when you model slightly more complicated things such as ring roads with queues at each roundabout. None of this takes account of the possibility that you are the person who gets stuck at the junction, triggering gridlock and a whole new multiplier effect.

6   The Highway Code figures for typical stopping distances are 96 m (24 car lengths) at 70 miles per hour and just 53 m (13 car lengths) at 50 miles per hour. The stopping distance has two components: the thinking distance, which is proportional to your speed, and the larger braking distance, which is proportional to the square of your speed. On this basis a lane at 50 miles per hour can take nearly twice the traffic of one at 70 miles per hour. So there is no need for anyone to queue when the lane closes, provided that no one leaves it to the last moment to change lanes. In reality most drivers don't leave as much as their stopping distance between them and the car in front, but the principles here still apply if they keep leaving the same proportion of that stopping distance between themselves and the next car as they slow down.

7   The carbon footprint tool is available as a free download from the Cumbria Tourism website, at <**http://www.cumbriatourism.org/sustainable- tourism/carbon-footprint-toolkit.aspx**>. It sets out to include just about everything that a business buys and does. For the most part it uses the same input–output model that I have drawn upon in this book. The calculator itself was put together on a limited budget but seems to work fine. It is designed for use by businesses of any size, and the ambitious intention is for this to be possible without businesses needing any external help. The tool was developed by my business, Small World Consulting. Thanks are due to Jessica Moss, who did most of the work, and to Cumbria Tourism and the Lake District National Park for funding assistance.

8   The data come mainly from Hammond, G. P. & Jones, C. I. (2008) *Inventory of Carbon and Energy (ICE) Version 1.6*. University of Bath. Available at <**http://www.organicexplorer.co.nz/site/organicexplore/files/ICE%20 Version%201.6a.pdf**>. See also Kalliala, E. M. & Nousiainen, P. (1999) Life

Cycle Assessment: environmental profile of cotton and polyester-cotton fabrics. *AUTEX Res J* **1** (1), 8–20. Available at <**http://www.autexrj.org/pdf/1999_No1/2.pdf**>.

9 Hammond & Jones (2008) (see note 8 above).

10 Estimates of the energy use per gigabit of transmission are 28 kg CO_2e over a UMST network and 31 kg CO_2e for GSM n. Based on Emmenegger, M. F. et al. (2004) Life Cycle Assessment of the mobile communication system UMTS: towards eco-efficient systems. *Int J Life Cycle Assessment* **11** (4), 265–276; and De Decker, K. (2008) 'The right to 35 mobiles', Low-tech Magazine, <**http://www.lowtechmagazine.com/2008/02/the-right-to-35.html**>

11 Including transport to the shop.

12 This is for a Nokia 7600, a fairly simple phone by today's standards. It has a camera and an MP3 player but it doesn't do e-mail or stop you from getting lost. Based on a Nokia life cycle analysis summarised in a WEEE Man Case Study Snapshot <**http://weeeman.org/html/what/lifecycle_case.html**> and referenced in Quaiguasi Froto Neto, J. (2008) 'Eco-efficient supply chains for electrical and electronic products', PhD thesis, Erasmus University Rotterdam, <**http://publishing.eur.nl/ir/repub/asset/14785/EPS2008152LIS9058921925Quariguasi.pdf**>. The figures are also broadly in line with those in Socolof, M. L., Cooper, D. & Dillon, P. (2007) *Expansion of the Electronics Environmental Benefits Calculator: Mobile Phone Reuse and Recycling.* Report submitted to Eastern Research Group, Lexington, MA. Available at <**http://www.abtassociates.com/reports/eebc_cellphone.pdf**>.

13 Compiled from data in the above four references

14 The estimate comes from De Decker (2008) (see note 12 above). This is also a good source of links for anyone digging around to understand the carbon impact of mobiles and other communications technology.

15 Emmenegger *et al.* (2004) (see note 12 above).

100 kilos *to* 1 tonne

1 A staggering 5 hours of life lost through death per 1000 miles of driving. My sum was just this: loss of life expectancy per mile = 2538 deaths on UK roads per year × 48 remaining years of life expectancy of an average driver, divided by 216 billion person car miles on UK roads per year = 5 hours life lost per 1000 miles of driving. (National Travel Survey, Department of Transport, 2009). I've based my sums on your having a life expectancy of another 48 years (I picked a 40-year-old man with a healthy lifestyle and because it gives me a nice round number), but you might want to adjust for your own situation. I haven't taken account of the fact that some of the deaths are of pedestrians (thinking that you might be just as bothered about

killing others as you are yourself) but I also haven't taken into account the possibility that you might acquire one of the 26,000 serious injuries or 150,000 minor injuries that are served up to UK car users each year. It's a lot better to be injured than killed on the road, but injury happens 10 times more often. I have also assumed that motorway journeys are averagely safe per mile compared with other car trips.

2   AA Routeplanner, <http://www.theaa.com/route-planner/index.jsp> [accessed 4 March 2009].

3   The figure of 20 per cent of presents being unwanted comes from Haq, G., Owen. A., Dawkins, E. & Barrett, J. (2007) *The Carbon Cost of Christmas*. Stockholm Environment Institute. Available at <**http://50plus.climatetalk. org.uk/downloads/CarbonCostofChristmas2007.pdf**>. The other numbers here are mine. This paper has an analysis in the same vein.

4   In Monty Python's *Meaning of Life*, Mr Creosote explodes in an unforgetable manner after being tempted into one last wafer-thin mint.

5   Energy Saving Trust is a body funded by the UK government that offers advice and grants: <**www.energysavingtrust.org.uk**>. Their sums are for a three-bedroom house and include reduced savings for 'comfort uptake': turning up the temperature a bit when the new insulation is fitted and cashing in on the added comfort that becomes possible. I have adjusted their figures slightly by adding 10 per cent to the carbon savings to take account of the emissions involved in supplying gas to your house as well as your burning it. I have also applied discount rates of 10 per cent per year to the financial savings because this gives a more realistic picture (see Discount rates, page 134).

6   The ICE database gives a figure of 1.2 kg CO_2e per kilo for rockwool. I've gone with these despite the problems that process life cycle analysis has with underestimating absolute numbers. Hammond, G. P. & Jones, C. I. (2008) *Inventory of Carbon and Energy (ICE) Version 1.6*. University of Bath. Available at <**http://www.organicexplorer.co.nz/site/organicexplore/files/ ICE%20Version%201.6a.pdf**>. I have allowed a density of 18 kg per cubic metre and assumed a loft area of 65 square metres. My figure probably undercooks it a bit, but the embodied emissions are still going to be a small deal in the overall sums.

7   I have assumed that the carbon footprint of extracting the gold has been 'written off' by previous owners.

8   This seems a reasonable ballpark figure based on data in Höhne, N., Phylipsen, D. & Moltmann, S. (2007) *Factors Underpinning Future Action: 2007 Update*. A report by Ecofys for the Department for Environment, Food and Rural Affairs. Ecofys GmbH, Cologne. Available at <**http://www.fiacc. net/data/fufa2.pdf**>, in which, for example, greenhouse gas emissions per GDP for China are 2.3 times greater than those of the UK and carbon emissions per tonne of steel are twice as high as in the UK.

9   Energy Star, sponsored by the US Environment Protection Agency, has downloadable spreadsheets showing typical energy consumption figures for all qualifying computers: <http://www.energystar.gov> [accessed October 2009].

10 Apple and the Environment, <http://images.apple.com/environment/reports/docs/iMac_21_5_inch_Environmental_Report_2009.pdf> [accessed February 2010].

11 Based on attributing the footprint of manufacture, transport and facilities across their product range, with computers accounting for $14.3 billion of a $32.8 billion revenue. Sales data (2008) from Apple Watch, <http://blogs.eweek.com/applewatch/content/corporate/apple_fiscal_2008_by_the_numbers.html> [accessed October 2009].

12 The difference is also about what I would have expected. A report by Small World Consulting and the Crichton Carbon Centre, 'The implications of truncation error in process-based lifecycle analyses of traditional buildings and their components' (Historic Scotland 2009), summarises various academic studies from around the world, all concluding that in the construction industry, process-based life cycle analysis typically succeeds in capturing only just over half of the total footprint of buildings. On the simple basis that computers are more complex than buildings, we might expect this 'truncation error' to be even larger still in computer manufacture. In this way the data from Apple reaffirm the 560 kg result from the input–output analysis.

13 De Decker, K. (2009) 'The monster footprint of digital technology', Low-tech Magazine, cites various life cycle analysis studies. Technology changes fast. It's not clear whether the capability of a machine is going up faster than the efficiency of production. In 2002 a 2 g chip could hold 32 megabytes of memory and had an estimated footprint of 38 kilowatt-hours of energy.

14 It draws 37 watts when idle, 2 watts in sleep mode and less than 1 watt when turned off.

1 tonne *to* 10 tonnes

1   Reported in the *Guardian*: 'Online government reveals NHS price list' (6 February 2004), citing UK government figures.

2   36.5p per kWh for new build. 41p for retrofit. See the UK government's Department for Energy and Climate Change website, <http://www.decc.gov.uk/en/content/cms/news/pn10_010/pn10_010.aspx>.

3   Chris Goodall, author of the excellent and acclaimed *How to Live a Low-Carbon Life* (published by Earthscan in 2007; updated in 2010) also runs a very good website, Carbon Commentary (see, for example, <http://www.carboncommentary.com/2009/07/15/686#more-686> [accessed October 2009]). He's done sums on the financial payback from micro-renewables.

His numbers look at least as good as anyone else's.

4 I've taken account of the fading efficiency of the cells and also the probability that by the end of the 40 years even our dirtiest electricity will not be coming from a friendlier fossil fuel than coal.

5 The figures are derived by using Defra conversion factors and their suggested 1.9 emissions weighting factor. I have not added on their 9 per cent uplift factor to take account of planes not taking the most direct route – so this is a 'best route' scenario. The factors I have used do take into account fuel supply chains. This makes a difference of just a few per cent. The embodied emissions in the plane and the footprint of airport infrastructure are not significant compared with the huge fuel burn of the jet engines. (So don't take airports too seriously if they tell you how carefully they are managing the carbon footprint of the airport building.) I reach similar figures running the model produced by David Parkinson and assuming a full flight. Overall this suggests that the numbers I am quoting are on the low side. The flight is unlikely to be full every time, and the Defra figures are optimistic for such long-range flights where so much of the take-off weight has to be fuel.

6 An emissions weighting factor of 1.9 can be inferred from the IPCC 4 assessment report, published by Cambridge University Press. This is also the figure suggested in Defra (2009) 'Guidelines to Defra/DECC's GHG conversion factors for company reporting' (<**http://www.defra.gov.uk/environment/business/reporting/pdf/20090928-guidelines-ghg-conversionfactors.pdf**>), Annex 6, footnote 10.

7 David J. C. MacKay's *Sustainable Energy Without the Hot Air* (2009), published by UIT Cambridge Ltd, (<**www.withouthotair.com**>) neatly explains the physics of this and many other carbon questions, including the case for electric cars over the internal combustion engine.

8 David Parkinson (2006) 'A new way forward for air traffic control' (Sensus Ltd), <**www.sensus-dp.demon.co.uk**>.

9 At constant altitude, a plane needs to have lift equal to its own weight. It takes more energy to gain altitude, but that is turned into potential energy that can be regained in descent provided that traffic control allows things to be done efficiently. David Parkinson of Sensus produced a model of fuel burn in flight that closely mirrors European Aviation Authority figures. It is great for looking at different scenarios of plane types and models over different distances and with different payloads. The most efficient distance for a 747 turns out to be somewhere around 3000 nautical miles (about 5600 km). On these short flights it is also possible to carry a substantial freight load, which is impossible on the longer-range flight because the plane would be too heavy to take off. If you are interested in David's model, contact him at <**newmodel@sensus-dp.demon.co.uk**>.

10 Ammonium nitrate (NH_4NO_3) fertiliser is 35 per cent nitrogen by weight.

The nitrous oxide (N_2O) that is released is 64 per cent nitrogen by weight. The 1 per cent of the nitrogen that is emitted is 0.55 per cent of the original weight of the fertiliser in nitrous oxide, with a global warming potential 300 times that weight in CO_2 equivalent. So 1–5 per cent nitrogen released to the atmosphere is 1.65–8.25 tonnes CO_2e per tonne of fertiliser applied to the crop.

11 All the agricultural data in this section came from a lecture by Professor David Powlson during a visit to Lancaster University in November 2009. He is working with the Chinese government to get the message across to farmers.

10 tonnes *to* 100 tonnes

1 The death also causes a carbon saving, which I have not factored in.

2 See note 2 for Chapter 8.

3 Rates vary for different renewables options. See the UK government's Department for Energy and Climate Change website, <http://www.decc.gov.uk/en/content/cms/news/pn10_010/pn10_010.aspx>.

4 'A study of the embodied energy of upgrading or replacement options for traditional buildings'. A report for Historic Scotland by Crichton Carbon Centre and Small World Consulting, October 2009. Available at <http://www.historic-scotland.gov.uk/index/learning/freepublications.htm>. The embodied emissions quoted here draw strongly upon input–output analysis. Process-based approaches were also used, giving figures about 40% lower. This is about as expected, given the systematic tendency for underestimation in process-based life cycle analysis.

100 tonnes *to* 1 million tonnes

1 Optimum Population Trust, 'Contraception is "greenest" technology' (9 September 2009), <http://www.optimumpopulation.org/releases/opt.release09Sep09.htm>.

2 The figures come from a study I was involved in for a pool in a town in Scotland in 2007. I can't say which because, even though I'm sure they wouldn't mind being named, I haven't asked them whether they would mind.

3 Worldwatch Institute (2009) *State of the World 2009*, 26th ed. (Earthscan, London), p. 32.

4 Mongabay.com, <www.mongabay.com> [accessed October 2009].

5 Amazon Fund, <http://www.amazonfund.org>.

6 Tollefson, J. (2009) Paying to save the rainforests. *Nature* **460**, 936–937.

7 Shuttle data from Wikipedia. Other figures in my calculations were: 31 MJ per kg for the solid fuel, 143 MJ per kg for the hydrogen. I used 0.07 kg

CO$_2$e per MJ as a general figure for emissions from the burning of fossil fuels and added 10 per cent for their supply chains up to the point of combustion.

8 'What do you care what other people think?' (Richard Feynman, 1989) is a fascinating and entertaining account of the technical and management failures behind the disaster. Also recommended for anyone who is trying to get some clear thinking into a bureaucracy.

9 4.6 tonnes per return trip if you live in Hong Kong. See Flying London to Hong Kong, page 135.

1 million tonnes and beyond

1 This estimate comes from the British Geological Survey (2005). The US Geological Survey estimates just 200 million tonnes.

2 If you've read the book from the start you will have gathered already that this list is just the easy bits and you could happily double the footprint if you were a bit more inclusive. It's best not to get too bothered on this occasion. The numbers come from *Feasibility Study for a Carbon Neutral 2010 FIFA World Cup in South Africa*, Department of Environmental Affairs and Tourism, Republic of South Africa, and Norwegian Embassy, 2009. Available at <http://www.norway.org.za/NR/rdonlyres/3E6BB1B1FD2743E58 F5B0BEFBAE7D958/114457/FeasibilityStudyforaCarbonNeutral2010FI-FAWorldCup.pdf>.

3 Ignoring the detail that the World Cup only goes on for a few weeks, and isn't even on all the time during that period.

4 The numbers are derived from Gartner estimates and UK Market Transformation Programme reports: *Case Study: EU Code of Conduct for Data Centres: Reducing the Energy Consumed by BT Data Centres (BT/Defra Pilot)* (February 2009) and *Global Carbon Impacts of Energy Using Products: Report for Defra/the Market Transformation Programme by Klinckenberg Consultants* (April 2009). Available from <www.mtprog.com> [accessed October 2009]. Various sources gave a picture consistent with my numbers, including <http://peakenergy.blogspot.com/2008/09/cutting-data-centre-energy-demand.html> and <http://news.cnet.com/Gartner-urges-action-on-data-center-emissions/2100-1022_3-6212965.html>

5 *The Rebound Effect: An Assessment of the Evidence for Economy-wide Energy Savings from Improved Energy Efficiency* (October 2007), UK Energy Research Centre. Available at <http://www.ukerc.ac.uk/Downloads/PDF/0 7/0710ReboundEffect/0710ReboundEffectReport.pdf>.

6 Worldwatch Institute (2009) *State of the World 2009*, 26th ed. (Earthscan, London).

7 Office of National Statistics (2009) 'UK environmental accounts: total greenhouse gas emissions by 93 economic sectors 1990 to 2007',

<www.statistics.gov.uk/statbase/ssdataset.asp?vlnk=5695>. This model underestimates the emissions from overseas because it is based on the assumption that overseas production is exactly as carbon intensive as the UK equivalent. We know this is flawed, because very often overseas industries are less energy efficient and also the electricity supply is often more carbon intensive. China is an obvious example where this is true.

Also included here in this analysis is 1.9 mark-up factor for aviation emissions to take account of the higher impact that high-altitude emissions are known to have. The pie also includes 'fixed capital formation': new buildings and other new infrastructure that although not 'consumed' are nevertheless something we continually demand.

8 Audsley, E., Brander, M., Chatterton, J., Murphy-Bokern, D., Webster C. & Williams, A. (2010) *How low can we go? An assessment of greenhouse gas emissions from UK food system and the scope for reduction by 2050.* WWF-UK.

9 This includes an emissions weighting factor of 1.9 for high-altitude emissions, as I explain on page 3.

10 Höhne, N., Phylipsen, D. & Moltmann, S. (2007) *Factors Underpinning Future Action: 2007 Update.* A report by Ecofys for the Department for Environment, Food and Rural Affairs. Ecofys GmbH, Cologne. Available at <http://www.fiacc.net/data/fufa2.pdf>, 'Country fact files'. The figures do not include international aviation and shipping. For the UK that would add about 10 per cent.

11 Jackson, T. (2009) *Prosperity without Growth: Economics for a Finite Planet.* Earthscan, London. A recommended read.

12 Kilotonnes of TNT equivalent.

13 Duncan Clark in <http://www.guardian.co.uk>, 'The carbon footprint of nuclear war' (2 January 2009), drawn from Jacobson, M. Z. (2009) Review of solutions to global warming, air pollution, and energy security. *Energy Envir Sci* 2, 148–173, DOI:10.1039/b809990c (first published as an Advance Article on the web on 1 December 2008: <www.stanford.edu/group/efmh/jacobson/PDF%20files/ReviewSolGW09.pdf>).

14 *The Three Trillion Dollar War* by Joseph Stiglitz, a Columbia University professor who won the Nobel Prize for Economics in 2001, and Linda Bilmes. See <http://www.democracynow.org/2008/2/29/exclusive_the_three_trillion_dollar_war>.

15 Using conversion factors of 0.33 and 0.22 kg CO_2e per pound sterling for UK output of defence and health services, respectively, and $1.6 per pound sterling averaged over the duration or the war so far. The figures assume that US and UK industries have the same carbon intensity. The model excludes direct emissions from combat itself. A large margin for error has been added in.

16 Worldwatch Institute (2009) *State of the World 2009*, 26th ed. (Earthscan, London): Chapter 6, pp. 56–58, 'Reducing black carbon'. Everything on black carbon is taken from this chapter.

17 Intergovernmental Panel on Climate Change (2007) *IPCC Fourth Assessment Report. Working Group I Report: The Physical Science Basis* (Cambridge University Press, Cambridge, Chapter 2; Ramanathan & Carmichael, op. cit. note 2.) Referenced in the Worldwatch Institute's piece on black carbon (see note 6 above). Radiative forcing from black carbon is put at 0.4–0.9 watts per square metre, in contrast with 1.6 watts per square metre for CO_2.

18 Intergovernmental Panel on Climate Change (2007) Global anthropogenic GHG emissions. In *Climate Change 2007: Synthesis Report* (IPPC, Geneva), p. 36. Adapted to include an emissions weighting factor of 1.9 for high-altitude emissions.

19 © SASI Group, University of Sheffield.

20 © SASI Group, University of Sheffield.

21 © SASI Group, University of Sheffield.

22 Earthtrends. World Resources Institute, Searchable Database, 2006 data. <http://earthtrends.wri.org/searchable_db/> [accessed October 2009].

More about food

1 Most of the figures here come from the input–output model and are in line with other estimates. Tim Jackson's paper for the Carbon Trust, 'The carbon emissions generated in all that we consume' (January 2006), <http://www.carbontrust.co.uk/Publications/pages/publicationdetail.aspx?id=CTC603>, includes estimates of the CO_2 emissions from cooking and storing food at home. See also 'Cooking up a storm' by Tara Garnett, Food Climate Research Network, University of Surrey (2008), <http://www.fcrn.org.uk/fcrnPublications/publications/PDFs/CuaS_Summary_web.pdf>.

2 Agricultural greenhouse gas emissions are from UK environmental accounts: Office of National Statistics (2009) 'UK environmental accounts: total greenhouse gas emissions by 93 economic sectors 1990 to 2007', <http://www.statistics.gov.uk/statbase/Expodata/Spreadsheets/D5695.xls> [accessed 25 January 2008].

3 *Guidelines to Defra/DECC's GHG Conversion Factors for Company Reporting* (2009). Produced by AEA for the Department of Energy and Climate Change (DECC) and the Department for Environment, Food and Rural Affairs (Defra).

4 From WRAP (2008) *The Food We Waste*. Waste & Resources Action Programme (WRAP), Banbury. Available at <http://www.wrap.org.uk/

downloads/The_Food_We_Waste_v2__2_.ec417f6e.5635.pdf>. The overall figure of 30 per cent waste includes bones and other bits that we don't all consider edible.

5   These include CO_2, which has a global warming potential (GWP) of 1, of course, but requires a high-pressure system.

6   See note 4 above.

7   Sources for this table and the sustainable fruit table: <http://sustnable. woodcraft.org.uk/1_act5a.htm>; Food Tourism Scotland, <http://www. foodtourismscotland.com/larder/seasonal.php>; Scotland's seasonal food, <http://www.scottishfoodinseason.com/index.cfm?page=5B0965F8-155D-131B-DEA25ADC9BD1F6E3&month=12>; What's in your plate – Scotland, <http://www.whatsonyourplate.co.uk/buy_in_season.html>. Thanks also go to the Crichton Carbon Centre for help in pulling this together.

Some more information

1   The two I'm most often asked about are Nigel Lawson's *An Appeal to Reason: A Cool Look at Global Warming* (Duckworth, London, 2008) and Channel 4's mischievous 2007 documentary *The Great Global Warming Swindle*.

2   Minx, J., Wiedmann, T., Barrett, J. & Suh, S. (2008) *Methods Review to Support the PAS for the Calculation of the Embodied Greenhouse Gas Emissions of Goods and Services*. A research report for the Department for Environment, Food and Rural Affairs by the Stockholm Environment Institute and the University of Minnesota. Defra, London. Available at <http://randd.defra.gov.uk/Document.aspx?Document=EV02074_7071_FRP.pdf>.

Index

Numbers in *italics* indicate figures; those in **bold** type indicate tables.

A

2K Manufacturing 81, 215
Africa: emissions 174
agriculture
   antibiotics in 73
   and deforestation 154
   methane emissions *see under*
      methane
   N_2O emissions 1
non-CO_2 emissions 172
air resistance 41, 118
air travel 59
   flying from London to Hong Kong
      return 135–7
   long-haul flights 137
   loss of life expectancy 119
   short-haul flights 137
   UK's footprint 164–5
   see also aviation emissions
air-conditioning 67, 109, 117, 161,
      168, 172
   non-CO_2 emissions 172
air-freighting
   air-freight labels 137
   asparagus 23–4, 83–4
   clothing 137
   food 137
   from developing countries 137
   roses 68, 214n31
   shoes 106
   strawberries 39
   vegetables 109

algae 89
aluminium 61, 62, 63, 181
Amazon 162
Amazon Fund 154, 192, 224n5
Amazon region 124
Americanos 36
ammonium nitrate (NH_4NO_3)
      fertiliser 224n10
animals
   animal welfare 97
   and plastics 18, 80
antibiotics in farming 73
Apple 125, 126, 127
apples 26–7
Aral Sea 94
asparagus 23–4, 83–4
Association of European Plastics
      Manufacturers (APME) 195
asthma 89
atomic war 169
Australia
   bushfires (2009) 162–3
   coal-dependency 168
   electricity 55, 56
   individual annual carbon footprint
      6, 139
   leaving the lights on 101
   target for cutting carbon
      emissions 7
aviation emissions 7, 8, 59, 117, 171,
      226n7
   in a 10–tonne lifestyle 7–8

emissions weighting factor
    195–6, 200
high-altitude 3–4, 136, 163n, 172,
    195, 205n3
measuring 3–4
model of fuel burn in flight 223n9
see also air travel

B

baby corn 83, 183
bags
    paper 21–2
    plastic 18–19, 21, 22
    reusable 19
bananas 23, 68, 182, 209n7
    environmental issues 28
    a fine low-carbon food xiv, 28
    nutrition 28
    over-ripe 29
    wasted 28
banks 128, 129
bathing 81–2, 215n6
battery farms 96
beef 28, 71, 86–7, 95–6, 109, 154, 179
beer 49–51, *50*, 110, 180
bicycles, banana-powered 117
biodiesel 89
biodiversity 154, 192
biofuels 89, 154
biomass 178
black carbon 163, 170–71, 170n
bobby beans 83, 215n7
books, paperback 75–6
Booths Supermarkets 72, 78, 177–8,
    178, 179, 180, 181, 196–7, 214n32
bottled water 12
'bottom-up' approach 194
Bowland Fresh 72
BP 176
Brazil: deforestation 154
bread 77–8, 77, 83
BREEAM (Building Research
    Establishment Environmental
    Assessment Method) 14
brine concentration 92

Brown, Derren 4
Building Research Establishment
    (BRE) 101
building societies 128
burgers 7–8, 23, 39, 86–7, **86**, 95
burial 115
buses 37–8, 66, 108
bushfires 162–3, 163n

C

cafés 110, 165
California: desalination 92
cappuccinos 35, 36
carbon
    black 163, 163n, 170–71
    costs xi, 227n16; n17
    defined 1
carbon calculator websites 2
carbon dioxide (CO_2) 1, 2
    duration in the atmosphere 205n2
    and farming 95, 178
    refrigeration systems 181–2
carbon dioxide equivalent (CO_2e)
    definition 2
    tonnes of *see* tonnes of carbon
carbon efficiency 166
carbon footprint
    abuse of the phrase 2
    average individual tonnes
        annually 6
    cars 143–5, *144*, **145**
    definition 1, 205n1
    direct vs. indirect emissions 3
    impossible to measure 4
    orders of magnitude 4–5
    'process-based' approach 143
carbon instinct xi, xiii
carbon intensity 168, 179, 180, 226n7
carbon neutrality *149*, 150, 225n2
Carbon Reduction Commitment
    207n5
carbon 'toe-prints' 2, 2, 4
card packaging 180
carpets 112–13, **112**
carrots 46–7

cars
    car crash 141–2, *142*
    carbon footprint 143–5, *144*, **145**
    commuting 7–8
    a congested commute 107–108, 219n4
    and cycling 24
    diesel fuel 87–8
    driving 1 mile 65–8, **68**
    emissions in UK 164–8
    fuel 59, 87–9, *88*
    London-Glasgow and back 117–19
    motoring costs 59
    a new car 143–5, *144*, **145**
    number of people in 42, 65, 66
    reducing carbon footprint 66–7, 191–2
    saving carbon without losing money 66–7, **67**
    true carbon footprint of 3
catalogues 44
catering industry 108–110, *109*
cattle grazing 87, 154
cement 74, 89, 90, 101
central heating 25, 54, 55
cereals with milk 23
cheese 106–7, 179
cheeseburgers 7–8, 23, 40, 86–7, **86**
chicken 179
children, carbon footprint of 151–2
China
    carbon emissions per tonne of steel 221n8
    cement 74
    coal-dependency 140, 167
    electricity 56
    emerging middle class 168, 191
    emissions 166, 167, 173, 174
    fossil fuel reserves 175–6
    greenhouse gas emissions per GDP 221n8
    individual annual carbon footprint 6, 139, 140
    nitrogen fertiliser 138–9
    steel manufacture 102
    textiles 112–13

choices, informed xii–xiii
Christmas excess 119–21, *120*
Clare, Dennis 170
Clarkson, Jeremy 108
climate change
    and black carbon 170
    caused by greenhouse gases 1–2
    dealing with xii, 190–91
    diesel pollution 89
    impact 2, 205n2
    importance of xii-xiii, 187
    man-made 187–9
    positive feedback loop 163
    and water stress 12
clothes rack 82
clothing 93–5, *93*, 137
Co-operative Bank 129
coaches 117, 118
coal
    China's dependency on 140, 167
    and electricity 56, 92, 113, 126
Code for Sustainable Homes Level 5, 150
coffee 35–7
cold storage 27
Compaq PCs 127
composting, and landfill 61–2, *61, 62*
computers
    and data centres 161–2
    emails 15–16
    the machine itself 125–6
    networks 124
    servers 124
    using 124, 126–7
    web searches 12–14
condensing driers 84–5, 215n9
construction industry 74, 149–50, 150n, 166
contrails 136n
cookers
    gas 24, 25, 55, 70
    pressure 71
cooling technologies 181
Copenhagen climate change conference (2009) 191

cost efficiency of selected carbon-saving options 191–92
cotton production 93, 217n22
cows, and methane 71, 72, 73, 87, 95, 178
Cranfield University 87, 96, 98, 196, 217n31
cremation 115–16
Crichton Carbon Centre 222n12
Crocs 105
cycling 23–4, 42, 132

D

dairy produce 53, 132, 179, 182
Danish Commission for Scientific Dishonesty 189
data centres 161–2
defence 165, 166, 221n11, 226n15
deforestation 28, 69, 70, 87, 154, 165, 177, 192, 209n8
Denmark: steel manufacture 103
Department for Environment, Food and Rural Affairs (Defra) 27, 194, 196, 223n4
desalination 12, 91–2, 216n16
desktops 124, 125
diamonds 123–4
diesel engines 171
diesel fuel 41, 87–8
    see also petrol
discount rate 121, 134, 148, 192
dishwashers 63–4
domestic energy 164
doors, walking through 14–15
driving 1 mile 65–7, **68**
drugs: health services 131
Dyson Airblade 17

E

Eco Kettle 26
Eco-Cement 74
Ecology Building Society 128–9
EcoSheet 81
ecosystems
    marine 92
    and plastic bags 18
education 166
eggs 96–7, *97*
electric hand driers 17
electric lights *see* light bulbs; lights
electrical goods 166
electricity
    Australian grid 50, 55, **56**
    and coal 56, 92, 113, 126
    computers 124, 126
    data centres 161–2
    from renewables 56–7
    generated from fossil fuels 56
    green tariffs issue 58
    health services *132*
    a 'high-grade' form of energy 55
    marginal demand **56**, 57
    nuclear 42
    a principal cause of carbon emissions 55
    refrigeration 181
    and steel manufacture 102
    trains 41
    UK grid 54, 55
    a unit of 55–58, **56**
electronic book readers 76
emails 15–16, 65, 127
emergency services 141
emissions
    agricultural 174, 178, 196
    average individual annual 6, 139–40, *140*
    aviation *see* aviation emissions
    biofuels 89
    bottled water 43–4
    bushfires 162–3
    car 107–8, 164–8
    computer 126–7
    direct vs. indirect 3
    electricity as one of the principal causes 55
    emissions per unit of GDP *167*, **168**
    fossil fuel 1
    global 9, 171–4, *172, 173, 174*

high-altitude *see under* aviation emissions
and mass annihilation 170
nitrous oxide 95
paper/plastic bags 21
volcanic 159
and water supply/treatments 12, 207–8*n*4
energy per mile 213*n*30
Energy Saving Trust 121, 221*n*5
Energy Star 222*n*9
energy transmission grids 55
energy-efficiency 4, 150
Europe: bottled water consumption 211*n*8
European Trading Scheme 207*n*5
Eurostar 42
exports 139, 140, 161, 167
Exxon 176

F

F gases 172, *172*
fair trade 28
fairy lights 119, 120, 121
family planning 152, 192
farms
    and food footprint 178–9
    see also agriculture
feed-in tariff 133–5, 147, 148, 192
fertilisers 89–90, 138–9, 174, 178, 224*n*10
financial services industry 128
fine art 60
Finlay, David 73
fish 99–100, *100*
'fixed capital formation' 226*n*7
flowers
    artificial 69
    cut flowers 68–9, 154, 182
    seasonal 68
food
    as 20 per cent of carbon footprint 165, 177
    and drink 165
    low-carbon food tips 182–3
    seasonal 47, 83, 182, 184–6
    see also food footprint
Food Climate Research Network 196
food footprint 177–82
    farm 178–9
    food waste 181
    hot houses 180
    meat and dairy 179
    packaging 180–81
    refrigeration 181–2
    transport 179
    waste 28, 78, 109–10, 119–20, 165, 177, 181
footprint: defined 1
fossil fuel reserves, burning 175–6, **175**
fossil fuels
    apple production 27
    diesel fuel 41
    electricity 55
    emissions 1, 87
    heating 54–5, 98
four-wheel drives 117, 119
France
    electricity 42, 64
    nuclear power 42, 101, 164, *166*, *167*, 168
fruit
    misshapen 27, 47, 183
    seasonal 110, 186
fuel
    combustion 171, 172
    diesel 41, 88–9, 171
    health services 131
fungicides 28

G

garden waste *61*
Gartner 161
gas boilers 25, 54, 55, 82
gas cookers 24, 25, 55, 70
    fan-assisted 71
geothermal energy 53, 56, 57
glass *61*, *62*, 71, 79, *79*, 180
gold 123, 124

Goodall, Chris 133, 146, 222–3*n*3
Google 12–14, 195, 208*n*6
green credentials 57–8, 212*n*15
greenhouse effect 136*n*, 159, 171
greenhouse gases
    'conversion factors' 88
    direct greenhouse gas emissions
        per GDP and per person for 60
        countries 197–9
    the dominant man-made
        greenhouse gas (CO_2) 1, 2
    see also carbon dioxide; methane;
        nitrous oxide
greenhouses 40, 68, 83, 98, 178, 180
Guardian newspaper 49, 115
Guardian Weekly 49

H

hand-drying xii, 17
handwashing dishes 63–4
Harrisson, John 74
health service 17, 131–2, 141, 226*n*15
healthcare 131–2, *132*, **169**
heart bypass operation 131–2
heat, a unit of 54–5
herbal tea 35, 36*n*
high-altitude emissions *see under*
    aviation emissions
hill farmers 111
Historic Scotland 149
homes, CO_2 emissions in 1
hospitals 165
hot-housing 28, 40, 180, 182
hotels 108–10, *109*, 165
house construction 149–50, *149*, 193
HP PCs 127
hydrocarbons 18, 80, 180
hydropower 56

I

ice cover, global reduction in 171
ice creams 53–4
ice lollies 53

Iceland
    clean energy 51–2, 53, 55–8, **56**, 57
    renewable-powered 168
iMac 125, 126, 127
imports 139, 140, 164, *165*, 167
incineration, rubbish 61
India
    emissions *167*, 174
    emissions per person 139, *140*
    steel manufacture 101
industrial processes: N_2O emissions
    1, 169
input-output model 105, 125, 126,
    146, 148, 169, 195–6, 219*n*7, 227*n*1
insulation 112, 221*n*5
    loft 121–3, **122**, 192
Intercity trains 40–42
Intergovernmental Panel on Climate
    Change 189
    2007 report 170
Iraq war 169–70
ironing 22–3, 64

J

Jackson, Tim: *Prosperity Without
    Growth* 61, 168
jeans 93–4, 140
jewellery 123–4
journalism 48
junk mail 44–5, 46, 128

K

Kemp, Professor Roger 41
Keswick Brewing Company 49, *50*
kettles
    electric 24, 25, 26
    gas 24, 25
kitchen waste 63
Kyoto Protocol 205*n*2

L

lamb 87, 99, 109, 111, 179
Lancaster University 156–8, *157*

Land Rover Discovery 65, 117, 143, 144, *145*
landfill sites
  anaerobic decomposition 45, 62
  and composting 61, *61*
  food waste 165, 189
  methane emissions 1, 21, 38, 45, 48, *48*, 71, 72, 171
  newspapers 48
  paper 21, 44–6, *45*, 180
  plastics 18, 80, 180
  and recycling 61, *61*, 62, *62*
  rubbish 61–3
landlines 114
laptops 127
lattes 35, 36
laundry 84–5, 93, 110
leather shoes 105–6
Leontief, Wassily 195
letters 44–6, *45*
life expectancy 119, 151, 220*n*1
light bulbs 7–8, 100–101, 177
  incandescent 100–101
  LED 120
  low-energy 101
lights: leaving the lights on 100–101, 177
limestone 74
line-drying 85
loft insulation 121–3, **122**, 192
logging 154
Lomborg, Bjørn: *The Skeptical Environmentalist* 188
London Underground 42
loss of life expectancy per mile 220*n*1
loss of person-years 142, *142*

M

McAfee 15
macroeconomic modelling 195
magnesite 74
Mailing Preference Service 46
mailshots 46
Malawi
  individual annual carbon footprint 6, 139

low footprint 168
manure 138, 172
Marcos, Imelda 105, 218*n*1
Marine Conservation Society 100
marine ecosystems 92
Market Transformation Programme 195
Marks & Spencer 77
meat in diet 95, 132, 154, 179, 182
mercury 115
Merz, Sinclair Knight 148*n*
metals
  car parts 143
  recycling 62
methane (CH_4) 1, 2
  coal, gas and oil processing 171
  cows 71, 72, 73, 87, 95, 178
  duration in the atmosphere 205*n*2
  global emissions 171, *172*
  kitchen waste 63
  landfill sites 1, 21, 38, 44, 47, *48*, 61, 80, 171, 180
  livestock 171
  rice cultivation 89-90, 91, 171
  sheep 111, 178
  water treatment 171
microprocessors 126
microwave 52, 71,183
Middle East: desalination 92
Miliband, Ed 39
milk 71–3
  carbon footprint 35, 40, 96, 179
  cheese-making 106, 107
  in porridge 51, 52
  in tea/coffee 35–6
milkmen 71
mining 154
misdirection of attention 4
mobile phones 1113–15, 114, 220*n*12, 220*n*14
Morphy Richards Ecolectric Kettle 26
mortgages 127–9
Mount Etna 159
Mount Pinatubo, Philippines 159
mugs 37

N

nappies 38–9
national consumption 164–8, *165, 166, 167*
National Geographic 18
National Health Service 131, 132
necklaces 123–4
newspapers 47–9, *48*, 109, 110
Nigeria: steel manufacture 102
nitrogen 89, 95–6, 138, 139, 172, 224n10
nitrous oxide (N_2O) 1, 2, 95, 96, 136n, 138, 172, 224n10
    and farming 178
    global emissions 174, *174*
Nokia N7600 phone *114*, 220n12
North American emissions per person 139
Norwegian government 154
nuclear 'baseboard' 42
nuclear energy 57
numbers sources 193–203
    Booths supermarkets greenhouse gas footprint model 196–7
    carbon footprint of UK products and services from different industries per £ of value 200–203
    direct greenhouse gas emissions per GDP and per person for 60 countries 197–9
    environmental input-output analysis 195–6
    publicly available data sets drawn from process life cycle analyses 194–5
nylon 80, 93, 94, 112

O

offal 96
Office of National Statistics 196, 200
open fires 171
Optimum Population Trust 152, 192, 207n4

orange juice 29–30
oranges 29–30, 182
orders of magnitude 5
organic
    eggs 96, 97
    farming 73, 87
    tomatoes 97, 98
    wine 78
Organico 79

P

packaging
    biodegradable plastic 81
    bottled water 43
    food 180–81, 183
    recycling 79
    saving food wastage 40
paper
    packaging 180
    recycled 21, 22, 44, 45, 46, 62, 65, 75, 76
    virgin 21, 44, *45*, 48, *48*, 65, 75, 76
paper bags 21–2
paper industry 21, 44
paper towels 17
paperback books 75–6
Parkinson, David 223n5, 223n9
PAS 2050 standard 194
payback calculations 133–5, 146–8
pesticides 28
PET (poly(ethylene terephthalate)) 43, 80, 211n7
petrol 87–9, *88*
    see also diesel fuel
Peugeot 107 car 65–6
photovoltaic panels 133–5, *149*, 154, 193
picking battles xii–xiii, 36n, 119, 191
pigs 87
plastics 80–81, 180
    data sets of emissions factors 194
    plastic bags 18–19, 21, 22, 180
    recycling 62, 80–81
polypropylene 21, 80, 112
polytunnels 40
polyurethane 112

porridge 94–5, **94**
Portland cement 70
Portugal, and coming climate change 216*n*21
postage, postal system 44, *44*, 128
potatoes 51–2, **51**
power stations
    CO_2 emissions 1
    coal-fired 134, 168, 171
    and desalination 92
    loss of energy 55
    nuclear electricity 42
Powlson, Professor David 224*n*11
pregnancies 152
presents, Christmas 119, 120, 121
pressure cookers 71
printing 44–5, 75–6, 128
process life cycle analysis 194, 195, 196, 224*n*4
processed meat 96
Profile Books 76
public administration 166
pubs *109*, 110, 165

Q

queuing theory 107, 108, 219*n*5

R

rail travel
    carbon footprint 40–43, 66, 117, 118
    and cycling 24
    loss of life expectancy 119
    safety 41–42
rainforest 59, 69, 128, 154
reading 24, 75
rebound effect 16, 162
recycling
    glass 71, 80, 180
    and landfill 61, *61*, 62
    metals 62
    newspapers 47, 48, 109
    packaging 81
    paper 21, 22, 44, 45, 46, 62, 75, 76

plastics 62
    textiles 62
Rees, Lord Martin 205*n*1
refrigerant gases 1–2, 205*n*2
refrigeration
    and F gases 172
    fish 100
    food 181–2
    ice creams 53–4
    non-CO_2 emissions 172
    orange juice 29–30
regenerative braking 41
Renewable Obligation Certificates (ROCs) 58
restaurants 110
reverse osmosis 91, 92
rice cultivation 89–91, *90*, 171, **216**
rockwool 221*n*6
rolling resistance 41
roses 68–9, 214*n*31
Royal Mail 46
rubbish 61–3, *61*, *62*
rucksacks 19
Rumencin 73
Russia
    coal-fired power stations 168
    cold climate 168
    factories 168
    fossil fuel reserves 175–6

S

salads 184–6
sales literature 45
saucepans
    boiling water 24, 25
    non-stick 52
Saudi Arabia 175–6
'save the planet' xii, 205*n*1
schools 166
Science Museum, London 81
sea burial 115
Seawater Greenhouse 92
sheep 111, 178
Shell 175–6
shoes 105–6, 218*n*1

showers 52–3, 82
silage 96, 178
Small World Consulting Ltd 193, 196, 200, 219n7, 222n12
social services 166
solar heating 54, 81
solar panels 59, 133–5
solar powered desalination 92
solar roofs 128, 145
South America: emissions 174
soya 154
    milk 73
Spain
    and coming climate change 216n17
    desalination 92
spam emails 15–16, 65
spending £1 59–61
standby mode 32–3
steak 95–6
steel 101–2, **103**, 181
strawberries 39–40, 109, 182
Stuart, Tristan: *Waste* 77–8
sugar: tea/coffee with no sugar *36*
Sun newspaper 47
supermarkets
    air-freight labels 137
    food waste 78
    milk 73
    plastic bags 18
    and refrigeration 181
    see also Booths Supermarkets; Tesco
supply chains 27
    banana 209n8
    cement industry 74
    fuel 38, 88, 117, 223n5
    and misshapen fruit/veg 27, 47
surgery
    and a car crash 141
    heart bypass 131
sustainable lifestyle 77
swimming 153
swimming pools 82, 152–3, *153*

T

tap water 12, 43
tea 35–7, 52
television watching 22, 30–33, 161, 210n10
    should you replace your TV? 31–2, 32
    standby 32–3
Tesco 29–30, 65
text messages 11, 114
textiles 94, 112
    recycling 62
thermal desalination 92
thermostats 26, 177
toilet rolls 64–5
tomatoes 97–8, 180, 182, 183
tonnes of carbon
    the 10-tonne lifestyle 7–8
    150 tonnes per life 8–9, 205–7n4
    financial value of saving a tonne of carbon 9–10, 207n5
    individual production 6
    what a tonne of CO_2e looks like 6
'top-down' approach 195
tourism businesses 110, 219n7
towels 109, 217n20
toxins
    and burning 18
    cotton production 95
trams 40
transport
    and food footprint 177, 179
    see also air travel; buses; cars; cycling; rail travel
trousers 93, 94
trout 99–100, **99**
truncation error 125, 222n12
tumble dryers 22, 84
    washing nappies 38, 39
2K Manufacturing 81

U

United Kingdom
  carbon efficiency 168
  carbon emissions per tonne of
    steel 221n8
  carbon footprint 164–7, *165*
  carbon footprint of UK products
    and services from different
    industries per £ of value 200–203
  construction industry 74, 150n
  electricity 55
  greenhouse gas emissions per
    GDP 221n8
  healthcare 131
  incandescent light bulbs 100
  individual annual carbon footprint
    6, 139–40, **140**, 164
  life expectancy 151
  steel manufacture 102
  target for cutting carbon emissions
    7, 151–2
  television viewing levels 210n10
  value of the UK population 9
United Nations 189
United States
  bottled water consumption 211n7
  carbon efficiency 168
  emissions 167, 173
  fossil fuel reserves 176
  individual annual carbon
    footprint 6
  steel manufacture 102
  target for cutting carbon
    emissions 7
units
  of electricity 55–9, **56**
  of heat 54–5
University of Bath: Inventory of
  Carbon and Energy 195
University of East Anglia 190
urbanisation 154
US Food Safety and Inspection
  Service 106
US Geological Survey 225n1

V

vegetables
  misshapen 27, 47, 183
  seasonal 47, 110, 182, 183–6
veggieburgers 86, 87
video conferencing 137
volcanoes 159

W

walking 42, 132
'wantedness factor' 119, 120
war-and-carbon discussion 170
washer-driers 84
washing machines 85, 140
washing powders 84
washing up 63–4
waste disposal centres 31
waste treatment 172
water
  boiling 24–6, 36–7, 183
  bottled 12, 43–4
  desalination 91–2
  electric water heating 81
  tap 12, 43
  water stress 12
  water supply/treatments and
    emissions 12, 171, 207–8n4
web searches 12–14, 127
Welsh gold 123, 124
wheat 90
wheelie baskets 19
wind farms 60, 128, 147–8, 192
wind turbines 145, 146–8, 192
wine 78–80, 79, 180
World Cup (South Africa, 2010)
  160–61, *160*
World Flowers 214n31
Worldwatch Institute 65, 170

And finally ... this full stop:

.

It's a particularly large full stop. I estimate that at 2 microns thick and 1 millimetre wide, it weighs about one five-hundredth of a milligram. At perhaps 10 kg per kilo for the ink that's a footprint of one-fiftieth of a milligram of CO_2e.

If just a few readers of this book have spent just a few seconds in quiet, low-carbon, contemplation of this black dot then it will have paid back its impact many more times over than the world's best offshore wind farms can ever hope to achieve. If it has distracted just a few people for just a few seconds from their shopping sprees, ski holidays, car journeys and Peruvian asparagus then, for its size, it will have made a truly outstanding contribution to the low-carbon world. An inspiration to us all.

fold

Mailing Preference Service

Freepost Lon20771

London WE1 0ZT

cut

The following people would like to opt out of all commercial mailing lists:

Your address	All names at the address